STRIKING CLOCK REPAIR GUIDE

STEVEN G. CONOVER

CLOCKMAKERS NEWSLETTER, READING

Dedicated to my wife, Karen

Published by Clockmakers Newsletter
203 John Glenn Avenue
Reading, PA 19607

Second Printing May 1996

Printed in the USA.

ISBN 978-0-9624766-4-8

CONTENTS

Introduction....iv

Count Wheel Movements

1 Setting up the Count Wheel Strike....1
2 Ingraham....5
3 New Haven....9
4 Gilbert
    Hour and Half Hour Strike on Gong....13
    Hour Gong with Half Hour Passing Strike on Bell....15
    Bim-Bam Strike on Chime Rods....17
5 Seth Thomas
    No. 89 Strike....18
    O.G. Strike with Alarm....21
6 Ansonia....25
7 Sessions....29
8 Hamburg-American....31
9 French Count Wheel....35

Rack & Snail Movements

10 Setting up the Rack & Snail Strike....40
11 English Bell Strike....45
12 French Rack Strike....52
13 Seth Thomas No. 120....56
14 New Haven....60
15 Hermle Model 130-020....64

Ship's Bell Movements

16 Ship's Bell Basics....67
17 Seth Thomas No. 7....69
18 Hermle Model 132-071....74
19 Schatz....78
20 Salem (Mauthe)....84

Appendix
Two Tools for Striking Clocks....88

INTRODUCTION

This book is a companion to *Chime Clock Repair*, first published in 1990. I am pleased to report that the chime book remains popular and has taken its place as a reference book on clocks which play Westminster or other chime melodies at each quarter hour. Many readers have written or called to say that the book has helped them to conquer repair problems in the eighteen featured movements.

Here, as in the earlier book, the idea is to help the repairer by providing illustrated procedures for the repair and adjustment of specific movements. This time the focus is on clocks which strike the hour or hour and half hour. The book will help all clock repairers, but there is even more here for the *beginner*, who is advised to work on striking clocks before attempting the more complicated chime clocks. The *professional* and the *skilled amateur* will also welcome the specifics on these striking clocks which, although not rare, may not turn up in the shop every month.

Striking Clock Repair Guide is divided into three sections covering count wheel, rack & snail, and ship's bell movements. Each section leads off with a chapter which explains the basics of the category. For example, Chapter 1 presents a count wheel strike setup procedure that will help repairers to assemble any count wheel device. Specific movements are then covered in individual chapters. Mastering each of these movements will enhance the abilities of the clock repairer, whether amateur or professional.

The book is not an encyclopedia of the striking clock movements made by every manufacturer. It is instead a repair guide to be used at the workbench, providing detailed repair instructions for a group of carefully selected strike mechanisms. Some, like the Ingraham count wheel strike in Chapter 2, are relatively common mechanisms that have much in common with other American count wheel movements. Others, like the Seth Thomas No. 7 count wheel ship's bell movement in Chapter 17, are mass produced items which are, nonetheless, one-of-a-kind designs. The English bell (Chapter 11) and the round French movements (Chapters 9 and 12) are examples of "types" made by various individuals and factories. Taken as a whole, the group of movements in this book is representative of the striking clocks, both new and old, in use today.

Steven G. Conover
Reading, Pennsylvania, May 1995

1

SETTING UP THE COUNT WHEEL STRIKE

Many striking clocks utilize the count wheel, which is also called the count plate or locking plate. The count wheel has slots arranged around the rim to stop the strike mechanism after the correct number of strokes on the gong or bell. The count wheel mechanism has few advantages compared to the rack and snail strike. The disadvantages, however, are well known to repairers: count wheel mechanisms can be tricky to assemble and may not stay synchronized with the hour hand if the clock is disturbed.

Nonetheless, the American count wheel mechanisms, with their relatively large, easy to handle parts, are ideal clocks upon which to begin the study of striking. Without a doubt, more count wheel striking clocks were made in the U.S. during the late19th and early 20th centuries than at any other time, making them affordable and easy to find.

Despite some individual differences in design, the American count wheel clocks have many similarities and can be studied as a group. Let's go over a simplified procedure for assembly and adjustment. Details of specific movements are covered in later chapters.

Procedure for Setting Up the Strike

A point of caution is in order for those who are just getting started repairing clocks. It must be remembered that the mainsprings must be *safely let down* before a clock can be taken apart in the first place. The mainsprings can be retained in wire loops which will contain all the power and limit the eight-foot-long springs to a manageable size. Figure 1 shows a mainspring with a wire loop wrapped around it. With the loop in place, hold a smooth handled let-down key on the winding arbor (Figure 2). Twist the key slightly in the winding direction to release pressure on the ratchet assembly. Use a screwdriver blade to pivot the pawl (which clock repairers call the "click") away from the ratchet wheel. Slowly allow the let-down key to spin, releasing the spring pressure into the wire loop.

Let's now assume the clock has been disassembled and cleaned. One more detail, often overlooked, is worth looking at now. When the movement is still apart, take the time to inspect and re-

Fig. 1. Mainspring with a wire loop wrapped around it

Fig. 2. Place a let-down key on the winding arbor, then release the mainspring power. Shown: Sessions demonstration movement with clear front plate.

place the brass wire springs on the strike arbors. These are nothing more than lengths of brass spring wire secured to a lever, wound around the arbor, then fastened to a pillar or to the movement plate.

Wire springs assure that the lever will move in a positive manner under spring pressure. There is usually a spring for the hammer arbor, which you can retain if it looks undamaged. There is another spring for the count, drop, and locking levers, and often a third for the lifting, unlocking, and warning levers. These springs are often found to be damaged. Leaving that kinked, old spring in place is not worth the chance that it will break in a month or two, making it necessary to disassemble the movement again before replacing it. The appendix presents an easy method for making the brass wire springs in a professional manner.

Some movements are easier to assemble by placing all the arbors in the back plate, then adding the front plate on top. Other movements are easier to assemble into the front plate, with the back plate put on last. After some experience, you can decide which is better in each instance. As a general rule, it is better to assemble into the plate which has the movement pillars attached to it, because this makes it easier to fit the loop ends of the mainsprings over the pillars. Assemble all the arbors between the clock plates, with the nuts or pins installed in the ends of the pillars, to hold all the parts together. Do not wind the mainsprings!

Step 1. Synchronize Count & Drop Levers

The first parts to check are the count and drop levers. Figure 3 shows that the drop lever (right) is located in the cam slot at the same time the count lever is down in one of the deep slots in the count wheel. Notice that the count lever lines up with the center of the count wheel. If it does not line up, bend the count lever to correct it. This adjustment can only be done up to a point; if you can't arrive at

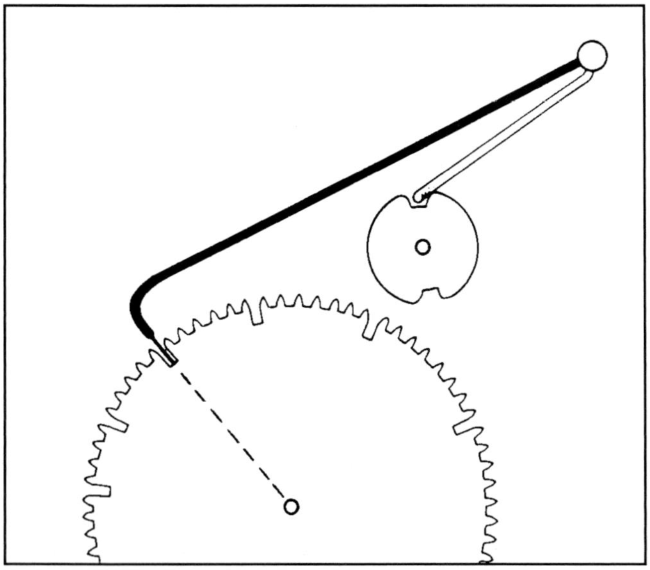

Fig. 3. Step 1: Synchronize the drop and count levers. The dotted line shows the angle of the count lever.

the appearance of Figure 3, you may have to disassemble the movement to change the mesh of the main wheel with the pinion above it.

If you take care while assembling the plates together, you can get the count and drop levers to stay close to their correct positions. Then all you have to do is bend the end of the count lever very slightly to adjust it. Remember the importance of having the end of the count lever fit exactly in the deep slots and point at the center of the winding arbor. That is, it must be radial to the count wheel.

Step 2. Adjust the Drop Lever

Now turn the wheels until the drop lever rides up on the cam, as shown in Figure 4. Check the amount of lift; you want to be sure that the count lever is held above the count wheel teeth as the clock

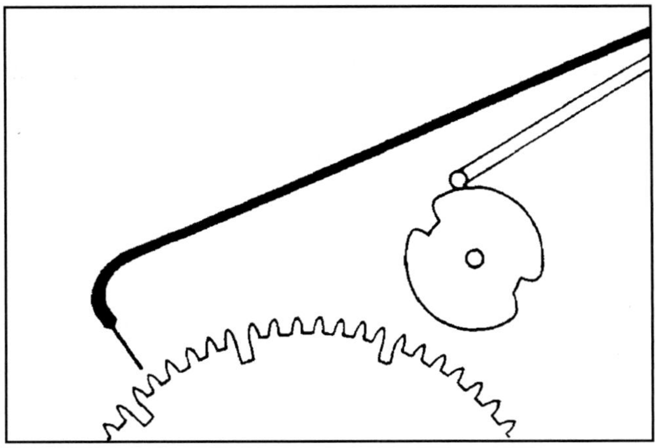

Fig. 4. Step 2: Adjust the drop lever.

strikes. To get more lift, bend the drop lever down. You could bend the drop lever up to reduce lift, but it usually isn't necessary to do this unless the drop lever appears to have been distorted by previous repair attempts. If you bend the drop lever, go back and check through Step 1 again. At this point, you have the strike set up so that the count and drop levers work well together.

Step 3. Set the Locking Lever

The locking action is set in this step. Bring the gear train around to the locking point. This means that the drop lever is in the slot of the cam and the count lever is in a deep slot of the count wheel, as shown in Figure 3. Now look for the lock pin. It should be next to the locking lever, which is the upper lever in Figure 5. If the pin isn't next to the lever, the strike train will not lock.

The adjustment is simple. Checking that the mainsprings are still completely let down, remove the pillar nut or pin at the upper strike area. Ease the plates apart enough to allow the locking wheel rear pivot out of its hole. (Some of the old movements were made with the pivot a little shorter, to make this easer to accomplish. But please don't use cutting pliers to snip off the pivot flush with the clock plate!) The pinion on the locking arbor will no

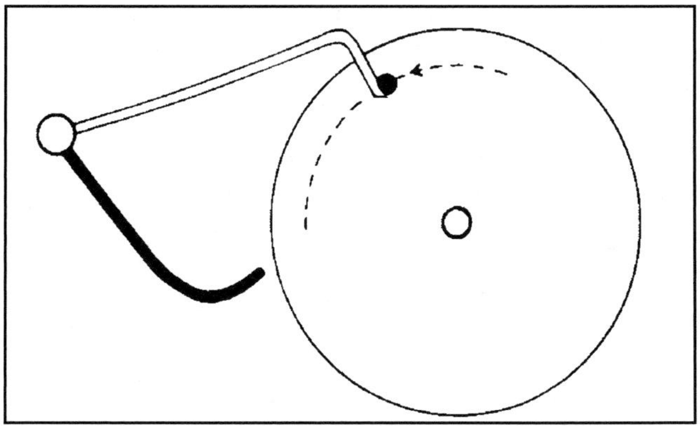

Fig. 5. Step 3: Set the locking lever.

longer be in mesh with the third wheel. Move the locking wheel, by itself, until the lock pin is at the locking lever. Re-mesh the gears again.

As part of this step, check two positions of the locking lever and lock pin. First, the locking position must be such that the lever completely captures the pin, with no worry of it failing to catch. Second, watch the lock pin as it passes the locking lever during striking. Each time the count lever goes into a shallow space between two teeth, the locking lever will move very close to the path of the lock pin. But we don't want it to lock in the middle of striking. Bend the locking lever up or down to achieve the best possible locking and clearance positions.

Now you have a movement that will count and lock. Fasten the plates securely together and wind the strike and time mainsprings.

Step 4. Adjust the Warning Action

Install the minute hand and turn it to about 10 minutes before the hour. As you continue to slowly move the hand, the clock will go suddenly to the

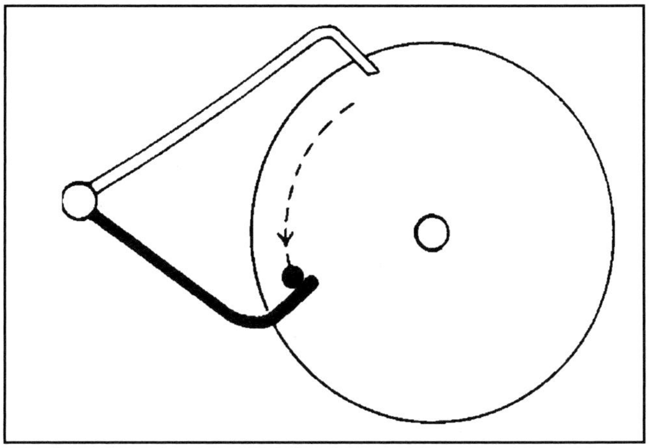

Fig. 6. Step 4: Adjust the warning action.

warning position between ten and five minutes before the hour. Figure 6 shows the warning lever drawn in black and the locking lever in outline. For best observation of the warning action, watch the clock run the last few minutes to reach the warning position under its own power. Turning the hands is too fast an operation, often hiding the fact that the warning lever does not catch the pin.

Adjustment is easily made to the warning lever. Bend the lever up to enable it to catch the lock pin. If, on the other hand, the lever still catches the pin while the clock tries to strike, bend it down until striking can work normally.

Step 5. Set the Warning Earlier or Later

Figure 7 shows the unlocking lever on the right. It lifts the count lever (at the arrow) to send the clock to the warning position. By bending the unlocking lever up, you can make the warning occur a

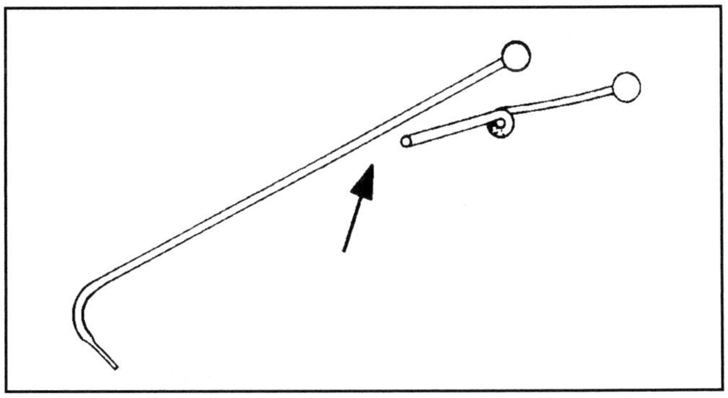

Fig. 7. Step 5: Adjust the time the warning occurs.

few minutes earlier. Bending it down delays the warning. Seven minutes of warning is average. Recheck the adjustment if the clock completely fails to strike; the unlocking lever may be too low.

Step 6. Set the Strike to Begin at the Hour

The star cam turns with the center arbor, and as it does so, it raises the lifting lever. When the minute hand reaches the hour marker, the cam has turned just far enough to drop the lever, and the strike begins. You may have to adjust the length of the lever, very slightly, by bending it and changing the shape. This will cause the lever to drop exactly at the hour. Many clocks that come in for repair begin to strike a minute or two before or after the hour; some are as much as five minutes off. Unfortunately, most American strikers don't have an adjustable minute hand bushing, which would have made the adjustment relatively easy to do. Some repairers mistakenly try to adjust the strike point by bending the minute hand, but the hand must be badly distorted just to change the strike point by one minute. The only result is a bent or broken minute hand. On this type of clock, the only effective way to adjust the exact moment the strike will begin is to bend the lifting lever.

On clocks which have a lifting lever made from soft steel wire, it is relatively easy to make the bend

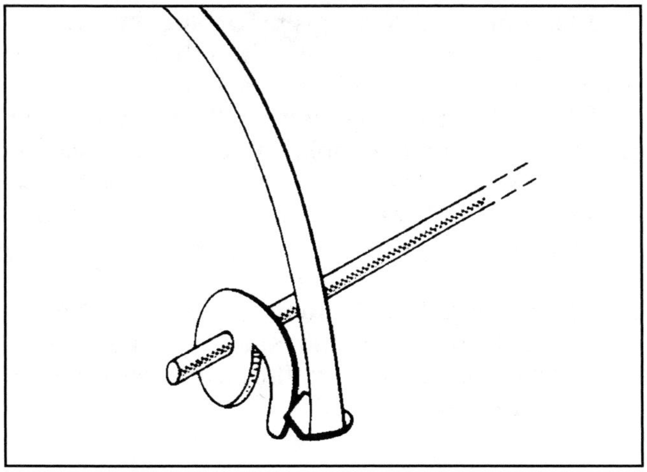

Fig. 8. Step 6: Set the strike point by bending the lifting lever. Do not break the lever at its tip!

required to set the strike point exactly at the hour. On the other hand, flat brass or steel levers such as the one pictured in Figure 8 may break if you bend them. Make the bend higher up, not on the sharp bend near the cam.

Hints on Adjustments

Adjustments to count wheel strike mechanisms can be frustrating. Make only one adjustment at a time, then observe the effect before going on to other adjustments. Trial and error adjusting can sometimes work, but not if you make a number of changes, one after the other, without stopping to assess the results. It is better to make a single adjustment with a clear understanding of what is intended.

Although soft steel levers can be bent over and over again, they can get distorted and marked up by pliers. The appendix describes a specially slotted screwdriver blade which can make accurate, clean bends in these steel levers. It is also important to look for deep grooves that may be worn into the drop lever and the tip of the count lever. These parts should be filed and polished smooth.

A Perspective on Count Wheel Strikers

The procedure described in this chapter is simplified and does not take into account several variations in count wheel strike mechanisms. For example, some New Haven movements lock directly on the cam and use the the pin we have called the "lock pin" only as a warning pin. There are many other differences in design, and this contributes a great deal toward keeping the work interesting.

The procedure explained in this chapter will, however, help you develop good adjustment skills on count wheel movements. It will also help you to break down a difficult job into smaller steps. The first time you encounter a mechanism someone has bent all out of adjustment, you will appreciate having a place to begin and a next step to follow. Adjusting a strike train which has had all the original lever settings ruined may take hours, so get comfortable and take your time!

2

INGRAHAM

Fig. 9. Ingraham "Nemo" mantel clock

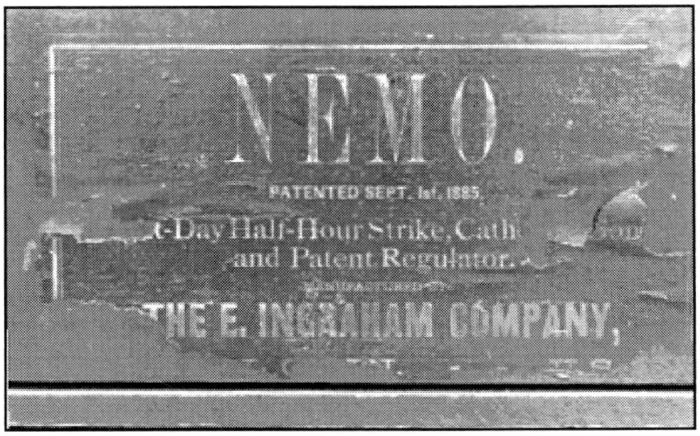

Fig. 10. The "Nemo" clock label

The Ingraham mantel clock shown in Figure 9 has a glossy black finish and a white dial imitating the appearance of certain French marble clocks of the late 19th century. However, the Ingraham case is wood, not marble; the dial is paper, not enamel; and the American movement is not much like that found in a French striking clock!

The original label on the case back (Figure 10) shows a patent date of Sept. 1, 1885 and describes the clock as an "eight-day half-hour strike, cathedral gong and patent regulator". Although the claim of "cathedral gong" sounds like typical advertising overstatement, the sound this clock produces is loud and resonant enough to be heard in another part of the house. Pencilled inside the case back is the repairer's notation "1-6-14 C.E.B. $1.00". Near the label is written "T.W. Sherman, Scobeyville" (N.J.).

The movement, shown in Figure 11, is stamped "E. Ingraham Co. Pat'd. Oct. 8, 1878, Nov. 11, 1879, Bristol, Conn." It is a common type from black mantel clocks of the late 19th and early 20th centuries, and it was used in other cases such as a plain varnished pine mantel clock and an oak school clock. The escape wheel is mounted on the outside of the movement, although it is not the type seen on the outside of the dial. The escapement is a half deadbeat, strip pallet type. A regulator device is mounted on the front of the movement, to the left of the escapement. On this model changes are made to the timekeeping rate by turning a knurled wheel which is accessible through a slot in the dial, just above the center arbor. Another version of this movement is regulated by turning an arbor located near the numeral "12".

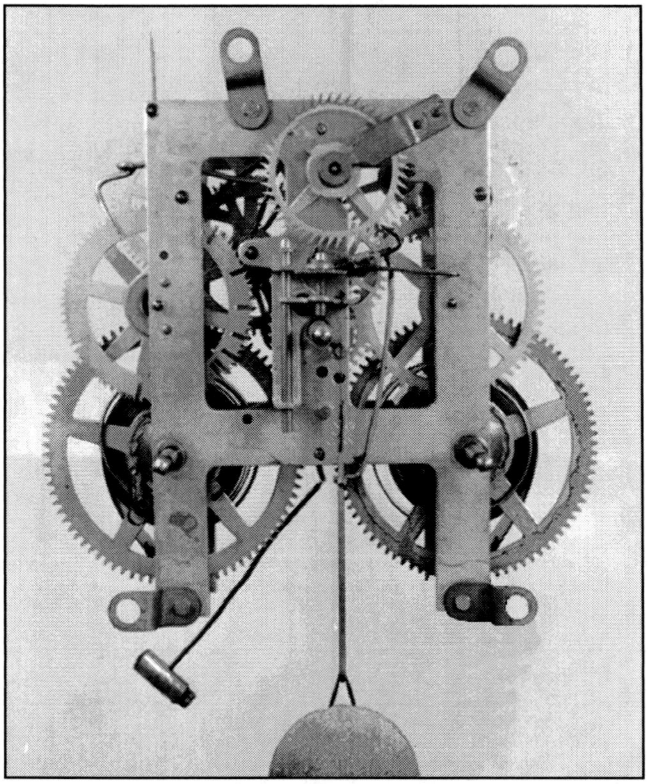

Fig. 11. The Ingraham strike movement

Design of the Strike Train

Hour striking is controlled by a count wheel, which is seen in the center of Figure 12. The count wheel is mounted at the second arbor location and is advanced at each hammer stroke by two pins extending from the lantern pinion shroud at the third arbor. This is in contrast to the more common setup, which is a count wheel mounted at the main wheel location and driven by a brass pinion on the second arbor.

The third wheel is a combined locking wheel and pin wheel, although we will refer to it as the locking wheel. This wheel has two lock pins and two hammer lifting pins on it. There is no cam on this arbor. Instead, the spaces between the locking pins allow the drop lever to descend, in much the same way as the slots in a cam would allow a drop lever to move down into them. This downward motion, combined with the count lever falling into one of the deep slots on the count wheel, allows the locking lever to catch one of the lock pins. This stops the strike train at the end of each hour count.

The fourth wheel is the warning wheel. Unlike most American count wheel mechanisms, the warning pin is just for warning, not locking. It must be set during assembly for a half revolution "run" at the warning period.

At the half hours, the strike hammer is lifted directly by the center cam, without any run of the strike gear train. This is called a "passing strike". In contrast, some count wheel clocks actually run at the half hour. It's easy to tell, in most cases, which type you have by checking the count wheel. If the strike train runs at the half hour, there is an extra single slot at the end of each hour segment on the count wheel, with 90 teeth in all. The Ingraham, with its passing strike, has only 78 teeth on the count wheel.

Assembling the Movement

Many American count wheel strike movements are mounted to the backboard of the case. The movements usually have the movement pillars riveted to the back plate, where the movement mounting tabs are located. This type of construction makes it better to assemble all the arbors into the back

Fig. 12. Closer view of the strike mechanism

plate, then add the front plate on top. Some movements, like this particular Ingraham model, are made the opposite way, with the pillars fastened to the front plate for mounting behind the dial. This is the common arrangement in black mantel clocks. It is better to assemble the parts into the front plate, then add the back plate on top. This allows you to place the mainspring loop ends over the lower pillars and hook on the hammer spring. Assembling "backwards" gives you the strike train view as drawn in Figure 13.

Following the cleaning of the movement and repair of worn parts, you are ready to reassemble the mechanism. Part names and numbers are those found on Figure 13.

The first step is to install the count wheel (18) in place and fasten it with the brass clip. To determine which way to install a count wheel, always look for the worn place around the center; this side goes against the brass disk riveted to the inside of the front plate at the second arbor location. Adjust the stiffness of the count wheel by slightly bending the brass clip. The wheel should turn without binding, but should be just tight enough that it always stays where it is turned. This kind of fit is required because of the drive method: two pins on the third arbor move the wheel in small "jumps", indexing it along one tooth space with each hammer blow. Other movements with the more common gear-driven count wheel could probably be set with less tension on the clip.

Add the escape wheel and its arbor to the front movement plate. Do this by sliding the escape wheel under the escape cock and easing the front pivot into its hole. The escape wheel cannot be added to the movement after the plates have been put together.

The sequence of assembly can be varied, but I have often just worked from the bottom of the gear trains upward. Install the center arbor (8) with hour wheel, time main wheel, strike main wheel (20), and the hammer (1). Add the time second wheel, strike second wheel (19), and the minute wheel, followed by the time third and fourth wheels.

Concentrating on the rest of the strike train now, install the locking wheel (9). Next, find the arbor with the lifting lever (3), warning lever (10), and unlocking lever (12). Install the arbor as shown in Figure 13. Add the other arbor, which has on it the count lever (17), locking lever (13) and drop lever (11). Do not attach the wire springs now, because they will cause the arbors to repeatedly pop out of position during assembly.

At this point, even before putting on the back plate, you can set up the strike train. One of the two lock pins (7) should be placed under the hook-shaped locking lever (13) as shown in the drawing. Set the count wheel (18) to permit the count lever (17) to go into one of the deep slots. Add the warning wheel (14), setting the pin about 180 degrees away from the warning lever. That is all there is to setting up the strike train, although you will almost certainly have to restore some of these settings after the plates are fitted together.

Add the fly (15) as the last item in the strike train. Then hook the wire spring for the hammer onto the center pillar. This is very difficult to do after the movement has been assembled. It is, however, relatively easy to do at any time before the plates are fastened together.

Place the back plate in position and get the rear pivots of the winding arbors in place first, followed by the hammer arbor. Proceed generally toward the top of the movement, placing each rear pivot into its hole. It is a good method to look for the pivot which seems to be holding the plates apart, then fit that particular pivot in place next. Add the lower pillar nuts (pins are used in this movement) to aid

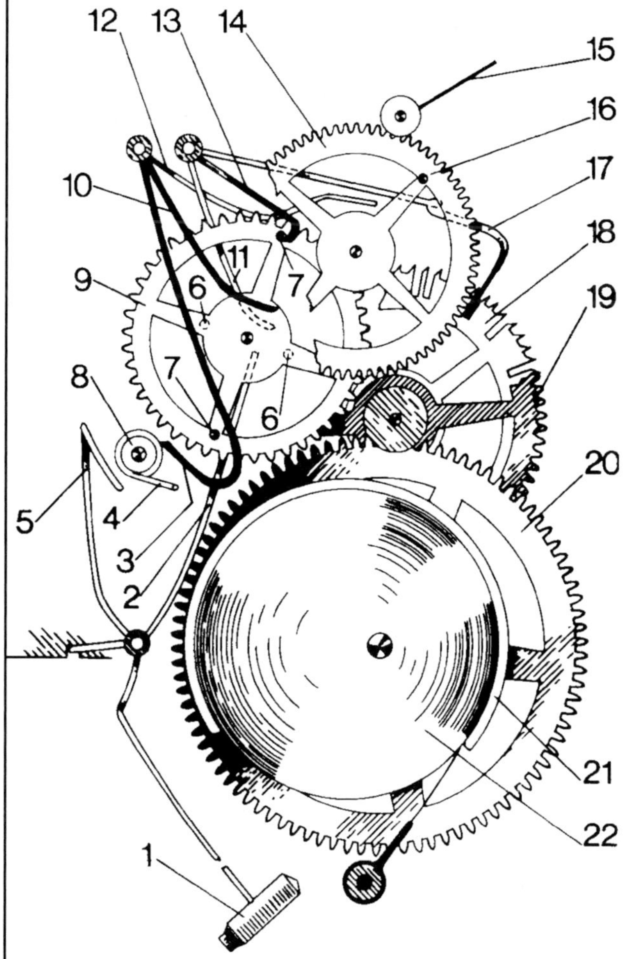

FIG. 13. INGRAHAM STRIKE MECHANISM, REAR VIEW

1 HAMMER	12 UNLOCKING LEVER
2 HAMMER TAIL	13 LOCKING LEVER
3 LIFTING LEVER	14 WARNING WHEEL
4 CENTER CAM	15 FLY
5 HALF HOUR STRIKE ARM	16 WARNING PIN
6 HAMMER LIFTING PINS (2)	17 COUNT LEVER
7 LOCK PINS (TWO)	18 COUNT WHEEL
8 CENTER ARBOR	19 STRIKE SECOND WHEEL
9 LOCKING WHEEL	20 STRIKE MAIN WHEEL
10 WARNING LEVER	21 MAINSPRING CLAMP
11 DROP LEVER	22 STRIKE MAINSPRING

in holding the work. After the plates are together, finish pinning or screwing the plates together. Finally, hook the two remaining wire springs for the strike arbors onto the pillar.

Set the Warning Run

The warning is the first thing to check after the plates are fastened together. Without being overly fussy, try for a 180 degree warning run. That is, with the gear train in the locked position, the warning pin (16) and the warning lever (10) are to be opposite each other. As warning occurs, the warning wheel and pin run about a half revolution until the pin is stopped by the warning lever. A run of less than 90 degrees (one quarter turn) may give a sluggish start-up action to the gear train. On the other hand, a warning run which approaches a full turn may result in interference with the warning lever.

To adjust the warning run, make sure both mainsprings are still completely let down. Remove the upper nuts or pins and separate the plates just enough to allow the removal of the warning wheel rear pivot from its hole. Disengage the pinion on the warning arbor from the third wheel, then make the adjustment. Verify that it is correct before going on.

It's worth noting that the warning run does not need to be set on most American count wheel movements; there is an automatic run from locking lever to warning lever. The Ingraham, however, requires that the warning be set, since the pin runs from a set position to the warning lever.

Other Adjustments

Fasten the plates securely, and partially wind the strike and time mainsprings. Install the minute hand, washer, and pin. Turn the hand very slowly toward the hour. For a true test, allow the movement to tick rapidly, with the pendulum off, rather than turning the hand to the warning position. At the instant the warning occurs, observe whether the warning lever safely catches the warning pin. If the movement begins to make a rapid ticking noise that is not from the escapement, check for a weak locking action at the warning. Bend the warning lever up for a deeper lock, down for a shallower lock.

Move the minute hand to the hour and start the strike mechanism. Observe whether the locking lever (13) remains clear of the lock pins (7) as they turn with the locking wheel. Then, at the locked position as the strike ends, note whether the locking lever embraces the full diameter of a lock pin. To adjust the depth of locking, bend the count lever (17) down to decrease the depth of locking, up to increase it.

Check the count lever. It must go cleanly into the deep slots in the count wheel. Notice how the count wheel moves in steps as it is alternately pushed ahead by the two pins extending from the locking wheel pinion. Bend the count lever slightly, as needed, to make the end point towards the center of the count wheel.

Turn the minute hand around to the warning point again. The unlocking lever (12) should push up the count lever to release the gear train for the warning. This should occur between ten and five minutes before the hour. Bend the unlocking lever up for an earlier warning, down for a later one.

With the pendulum off, allow the movement to tick rapidly to the hour without your touching the hand. Check to see if the strike begins with the hand pointing straight up. If the strike begins early or late (allowing some tolerance for the natural looseness in the motion work gearing) then you will have to adjust the lifting lever (3). Bend it at a point about 1/3 of the way up from its hook-shaped end. This point is not far from the rear pivot of the locking wheel. Bend the lever down to make the clock strike later, up to make it strike earlier.

Adjustment of the half hour passing strike is done independently of the hour. Using the same principle, however, bend the half hour strike arm (5) until the half hour strike occurs as nearly as possible on the 30-minute mark.

3

NEW HAVEN COUNT WHEEL STRIKE

In the late 19th and early 20th centuries New Haven Clock Co. made many clocks containing count wheel strike movements. The example shown in Figure 14 is the New Haven No.95, a black wooden mantel case with four columns applied to the front. The capitals and bases, as well as the feet and side

Fig. 14. New Haven No.95 mantel clock

ornaments, are cast metal. The columns are thin sheet metal. The case is 16-1/2 inches in overall width.

Figures 15 and 16 are photos of the front and rear of the movement. This common New Haven mechanism is a count wheel striker powered by loop-end mainsprings. The escape wheel and pallets are between the plates. The pendulum is mounted at the rear of the movement. Brackets are attached to the front of the movement to permit installation of the movement behind the dial. The hammer extends from the rear center of the movement out to the strike side, where it reaches the gong.

The count wheel is located directly in front of the strike main wheel and is driven by a gear on the second arbor. The deep slots in the count wheel are the end points of each strike counting sequence. On this count wheel the deep slots come in pairs, which indicates that the movement strikes both the hours and half hours by means of the gear train. In some other strike movements, the center cam merely lifts the hammer at the half hour for the strike, and the gear train runs only at the hour.

In this particular movement, the pillars are fastened to the rear plate. This makes it easier to assemble the parts into the rear plate, then add the front plate. Remember that both mainsprings must be fully let down into mainspring clamps before the plates are separated.

A noteworthy feature of the New Haven strike mechanism is the method of locking. Refer to Figure 17, on page 11, as the various strike parts are

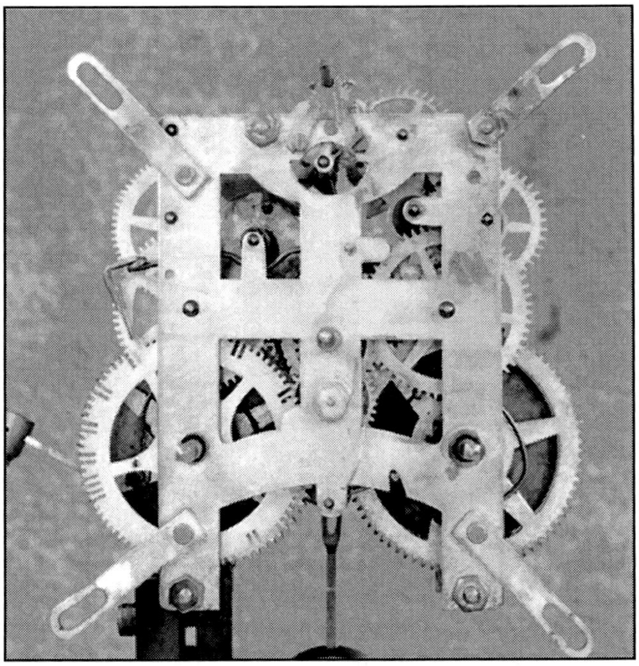

Fig. 15. Front view of the New Haven movement

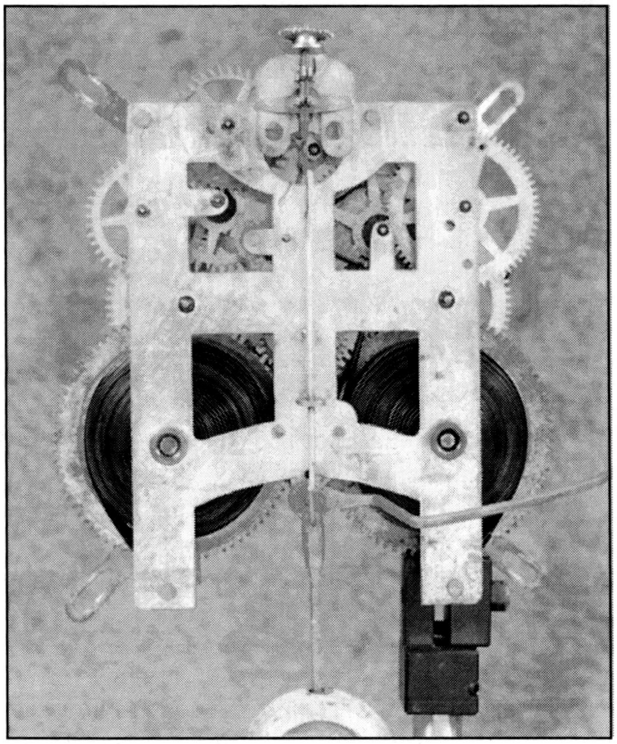

Fig. 16. Rear view of the New Haven movement

described. Locking occurs on the cam (6). The drop lever (7) doubles as a locking lever. There is no separate locking lever, and the warning pins (2) are for warning only, not locking. The drop lever rides on the cam, holding the count lever (3) out of the teeth in the count wheel (4). After each hammer blow, one of the two cam slots appears directly under the drop lever. This action permits the count lever to enter a deep slot in the count wheel if the strike count is finished.

Figure 17 shows the one-piece stamped steel lever which is fastened to a stud on the rear movement plate. The lift lever (11) is the lower part of this piece. It takes the lifting action from the center cam (9) before each hour and half hour. The other part of this steel piece is the warning lever (13), which actually has a double purpose.

In addition to its warning function, it raises the unlocking lever (10) as the clock moves toward warning. When the drop lever is lifted out of the cam slot, the movement goes to the warning position. The warning wheel (1) rotates about one quarter turn until one of the warning pins (2) is stopped by the warning lever (13).

The other set of strike levers is also shown in Figure 17. The count lever (3), drop lever (7), and unlocking lever (10) are all fastened to the same arbor. This means, of course, that all three move together. At warning, the unlocking lever raises the drop lever and count lever out of their respective slots. When the center cam releases the lift lever, the warning lever releases the warning pin. The gear train runs until the count lever drops into the next deep slot.

One of the major adjustment points in the New Haven strike movement is the cam (6) and its relation to the drop lever (7). A look at the cam reveals that there are two square slots, 180 degrees apart. These are notched from beveled-off portions on the cam. As a result, three different cam positions are created for the drop lever to ride upon— the slot, the bevel, and the rim. When the gear train is locked as in the Figure 17 rear movement view, the drop lever is in the slot, locked against the edge. At the same time, the count lever is in one of the deep slots in the count wheel. As the drop lever rides on the rim of the cam, the count lever is held high enough so the count wheel can rotate underneath it. In the third position, the drop lever is on a beveled portion of the cam. During the warning, the drop lever rests there after coming out of the slot. At each hammer blow, the drop lever returns to the bevel as the count lever drops into a shallow tooth space on the count wheel.

It is important to adjust the drop lever carefully. It must go deeply enough into the cam slot to lock on the edge of the slot. But the locking action must also be shallow enough so that the lever will stay out of the cam slots each time a slot passes under the lever during the strike cycle. In the locked position shown in Figure 17, front view, the locking action is safe, yet just over half of the diameter of the lever is within the cam notch. If the locking is made deeper, the lever may lock each time the cam revolves a half turn, no matter what the count wheel indicates. Another version of this problem occurs when the locking is only slightly too deep. As the clock strikes, you will hear a clicking sound as the drop lever glances off the edge of the slot following each hammer blow.

To adjust the depth of the drop lever, bend it toward or away from the cam slightly. Notice that the drop lever has a right angle bend near the end, and try not to distort its basic shape. Bending the count lever will also produce a corresponding change in the drop lever. However, you must remember to check that the count lever enters the deep slots as straight as possible. As shown in Figure 17, front view, the knife-edge of the count lever points at the center of the count wheel. When the drop lever rides on the rim of the cam, make sure the count lever is raised safely over the count wheel teeth.

Warning adjustment is rather simple. During assembly, orient the warning wheel as in Figure 17, rear view. This will permit a quarter turn warning run before the warning pin (2) hits the warning lever (13). Note that there are two warning pins, and

FIG. 17. FRONT AND REAR STRIKE TRAIN DRAWINGS

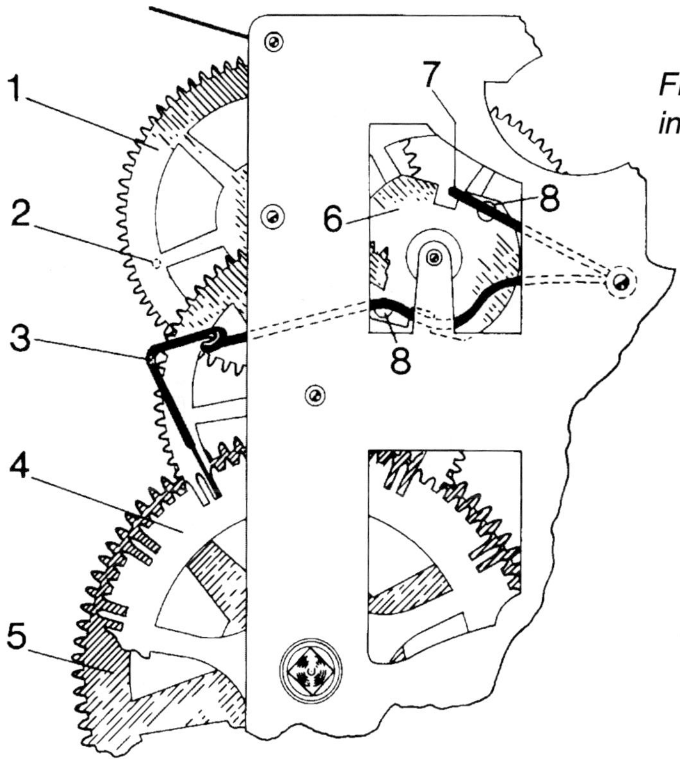

Front view, strike train in the locked position

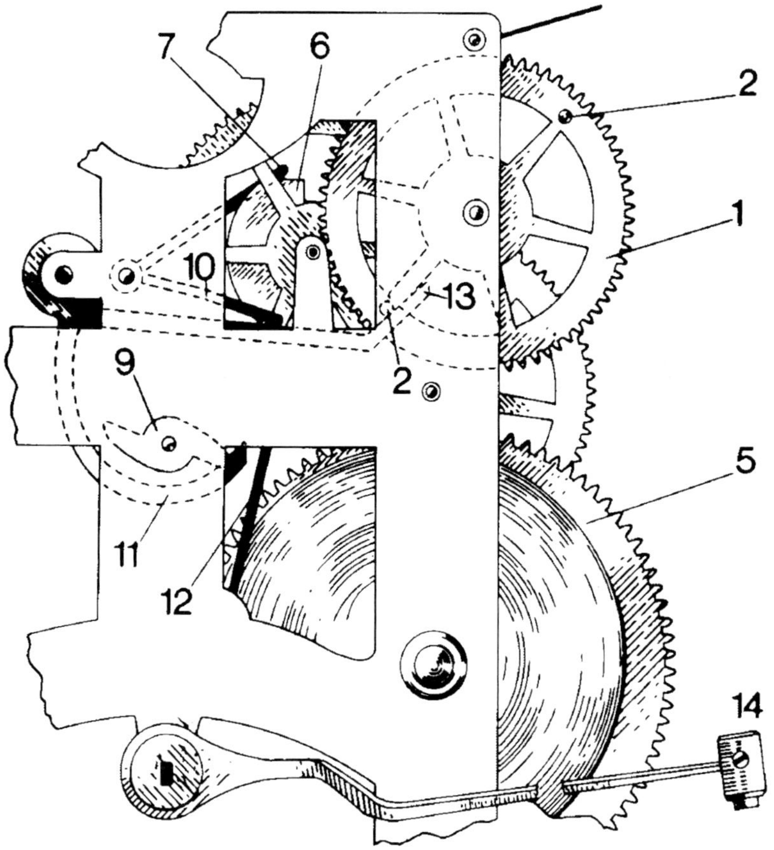

Rear view, strike train in the warning position

PARTS LIST

1 WARNING WHEEL
2 WARNING PIN (2)
3 COUNT LEVER
4 COUNT WHEEL
5 STRIKE MAIN WHEEL
6 CAM
7 DROP LEVER
8 HAMMER LIFTING PINS
9 CENTER CAM
10 UNLOCKING LEVER
11 LIFTING LEVER
12 HAMMER TAIL (PORTION)
13 WARNING LEVER
14 HAMMER

that for this reason you cannot have a half turn of warning. It should not be necessary to bend the warning lever up or down to adjust it. It is made from flat steel and is not as likely to have been bent by a previous repairer as would a lever made of soft rod. Just check the warning lever to see if it locks safely on the warning pins. If it is too close to the warning wheel or to the rear plate, straighten it.

The lifting lever (11) is part of the same piece of flat steel. A brass piece is riveted in the end to act as the pickup for the center cam (9). On many strike movements you will find that striking begins early or late, and that by bending the lift lever you can correct the problem. The New Haven lever does not lend itself to precise bending because of its construction. If you must bend it, take care not to loosen the brass piece in the end of the lever. The lifting and warning levers do not normally require a wire spring. They are heavily balanced in the downward position because of the long horizontal warning lever. The arbor which carries the drop, count, and unlocking levers should have a spring to insure positive locking.

Assembly and Testing

Assembly of the New Haven movement follows the general pattern of American striking clocks with the count wheel arrangement. The warning run, however, must be set as described earlier. This is because the warning pins do not work against a locking lever which would determine their position automatically. During assembly, try to put the strike train in the locked position. With the drop lever against the side of the slot, set the warning wheel for about a quarter turn of run. Place the count wheel so the count lever will go directly into a deep slot in the wheel. Some minor bending to correct the angle of entry for the count lever is usually necessary.

After assembly, check the operation of the strike train by applying finger pressure to turn the gears. You may have to ease the warning wheel rear pivot out of its hole to allow you to mesh the gear more accurately for the warning run. Always make sure there is no mainspring power on when you make the adjustment.

Hammer operation does not require much in the way of adjustment. The hammer tail (12) works in front of the cam, which has two hammer lifting pins (8) driven into it, one next to each slot. The location of the hammer lifting pins will assure that the hammer tail will always remain clear of the pins when the strike train is at rest. You may have to bend the hammer tail to the front or rear if it is either too close to the cam or in danger of missing the lift pins. A wire spring is required on the hammer arbor.

Testing of the New Haven strike movement involves a one or two week test run. If the strike train makes no counting errors and maintains its speed of striking fairly well over a seven or eight day period, no further adjustments should be needed.

4

Gilbert:

Hour and Half Hour Strike on Gong
Hour Gong with Half Hour Passing Strike on Bell
Bim-Bam Strike on Chime Rods

Many Gilbert mantel clocks made in the 1920's and 1930's are still in use today. Although some of the earlier Gilberts are collectors' items such as brass crystal regulators with glass sides and ornate china-cased clocks, many of the later Gilbert clocks are plain mantels that you wouldn't look for unless you were a real Gilbert fan.

Hour and Half Hour Gong Strike

Our first example for this article came from a Gilbert grandfather clock. I have repaired several of these clocks which had ladder chain movements and front mounted, half deadbeat escapements. This particular one was a spring driven model with a seconds-length pendulum. These grandfather clocks have a simple, early American appearance that fools some owners into thinking they have a much older antique. Once the dial comes off, the mantel-style movement is revealed. The movement shown in Figure 18 is stamped "34", which may be 1934. Repairers have said the movements are difficult to assemble and adjust. In this chapter we will look at three versions of the Gilbert strike mechanism to try to discover the reasons.

Figure 18 shows a front view of most of the strike parts plus two detail views of other parts. The center cam (10) has two equal arms, one for the hour and the other for the half hour. This tells us at once that the movement counts both the hour and half hour on the train. It doesn't have the passing strike at the half hour. One arbor holds the lifting lever (9), the warning lever (8), and the unlocking lever (4). The first two are stamped from a single piece of flat brass, which makes them hard to adjust and easy to break. Together, they unlock the strike train for the warning.

Another arbor carries the count lever (2), locking lever (3), the drop lever (5). You can expect to find adjustment problems here. Frustrated repairers have spent hours bending the levers before you. No matter how well you understand the movement, it can take more hours to unravel error upon error.

The cam (6) has two slots which allow the levers to drop into the locked position. It also has two stamped-out tabs to act as hammer lifting pins. One of the tabs is visible in the Figure 18 front view.

The Count Wheel. This Gilbert count wheel is different from most others in American clocks, for it has a smooth arc instead of individual teeth to mark each hammer stroke. In most count wheels, a deeply notched pair of teeth normally marks the end of the hour, then the half hour. Instead, the Gilbert has a plain slot which is as wide as the deep cut pair would have been.

This might not seem to make any difference as far as adjustments are concerned. In fact, it might seem easier to work with the wide slots in place of the narrow spaces between teeth on other count wheels. But this is not so; adjustments are highly critical. Although most count wheels have the hours one through 12, the Gilbert has one through 12 twice. (Not all Gilberts are made this way.)

The usual count wheel on a movement which counts both the hours and half hours on the gear train has 90 teeth. This 24-hour Gilbert count wheel would have twice as many, 180 teeth, but the teeth

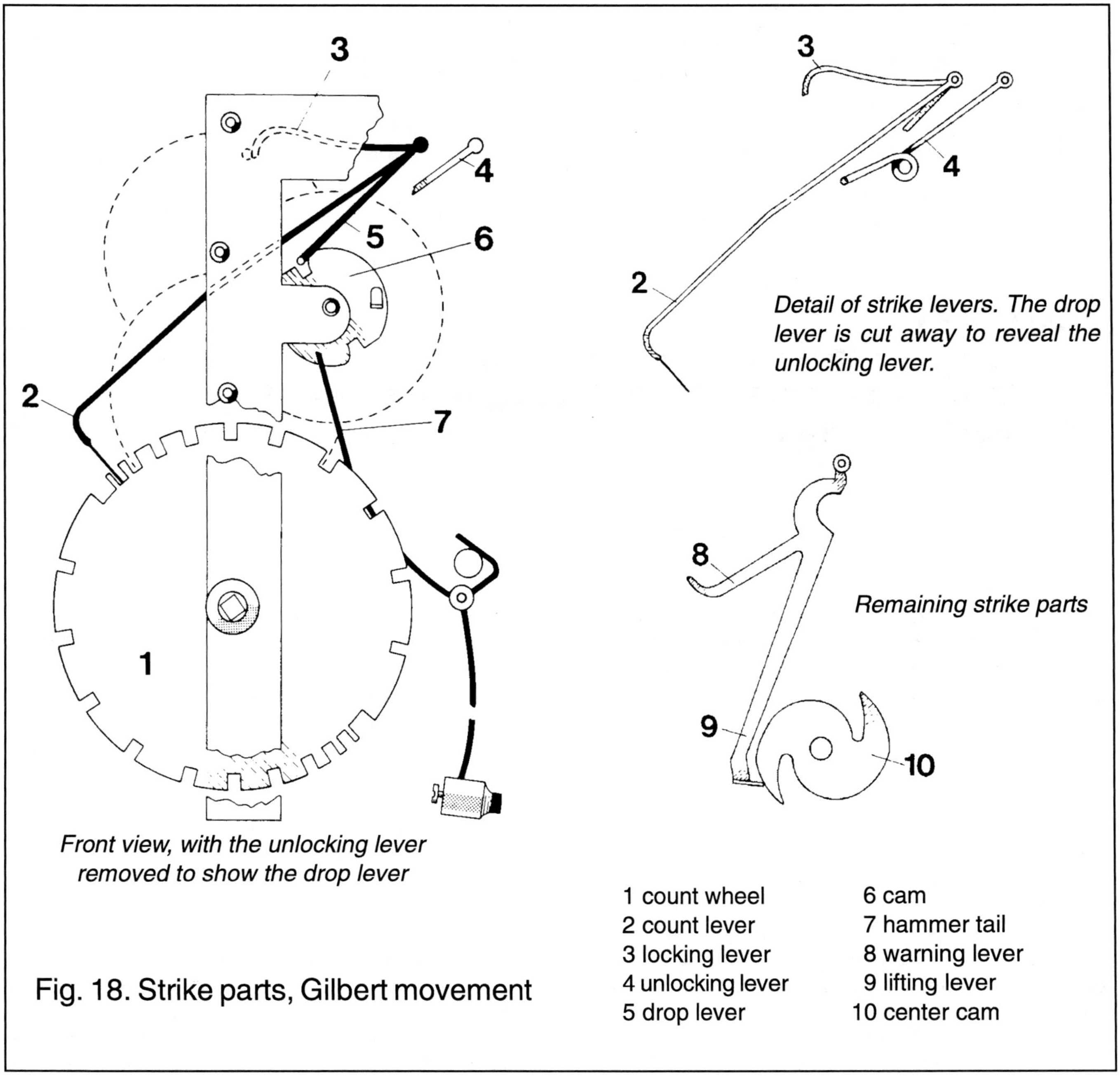

Fig. 18. Strike parts, Gilbert movement

aren't there. The count wheel is fastened to the main wheel with three brass pillars, and it moves with the wheel. That eliminates the need for teeth as a means of receiving power from a pinion on the second arbor. In addition, the teeth in a 24-hour count wheel would be too small and fragile: hence the smooth arcs and double-width slots.

Unfortunately, the adjustments are just as critical, even though the 180 fine teeth aren't there. At the end of the hour, the count lever must drop into the left side of the slot. Then after the half hour, it stops in the right side of the same slot.

The usual count lever adjustments are especially important here. The tip of the lever must point directly at the center of the count wheel circle. If it does not, it will enter the slots at an angle and will probably not clear them reliably. Counting errors will begin to show up.

Other Adjustments. It seems that the drop lever (5) is also hard to adjust. It should enter the cam slot and stop without touching either side. The slot is narrow, so you must make a careful adjustment. With the mainsprings completely let down, separate the plates enough to ease the rear pivot for the locking wheel out of the hole. Center the drop lever in the slot, then set the locking wheel so the pin rests on the locking lever (3). Figure 18 shows this position.

There is usually a hole in the rear plate which allows you to see the depth of locking between locking pin and lever. The best you'll get is a safe lock at the end of the strike and bare clearance when the train runs. With too little lock, the clock may continue striking extra hours. With too much, the clock will not strike at all. A rapid clicking noise during the strike means the lever glances off the pin at each revolution. Just set the lock shallower.

The Figure 18 detail view (top right) shows the unlocking lever (4). If you bend it up, the clock goes to warning earlier; bend it down and the warning occurs later. The warning should occur five to seven minutes before the hour and half hour.

The levers in the Figure 18 detail view (bottom right) should be approached carefully. If you don't absolutely have to adjust them, don't. The lifting lever (9) is brass in this model, and it may have been bent before to secure an earlier or later drop-off point for the strike. One bend too many means a break. The warning lever (8) should not need adjustment unless it bears the marks of earlier bending. You can check for a safe warning action by watching the movement as it runs. See whether the warning lever stops the pin safely at the instant the warning happens. If you turn the hands to check, you may move the levers too quickly, disguising a problem.

It's fair to say that this model Gilbert is relatively difficult to adjust, because of the design of the count wheel. There is no tolerance for error on adjusting the strike levers.

Hour Gong with Half Hour Bell Strike

A variation of the Gilbert strike mechanism is shown in Figures 19, 20 and 21. In this case the gear train runs only at the hour, as a hammer strikes the gong. At the half hour, a separate hammer tail is lifted and dropped by the cam, producing a single tone called the passing strike. This is often set up as a passing strike on a bell, which is common in black mantel clocks made around the turn of the 20th century. The passing strike was also set up to sound on the same gong used for the hours. Still another variation, illustrated at the end of this chapter, is a bim-bam strike on chime rods, with one note sounding at the half hour. These are all versions of the same thing: an hour striker with passing strike for the half hour.

Figure 19 shows the front of the movement. It appears quite similar to the movement discussed in the beginning of this chapter which features a smooth count wheel. However, the slots on the count wheel are narrower, so that only the hours are struck on the train. If there were a tooth on the count wheel for each note to be struck, there would be 156. That is, 12 + 11 + 10, etc. equals 78 strikes per hour, taken twice because there are two 12-hour segments on the count wheel.

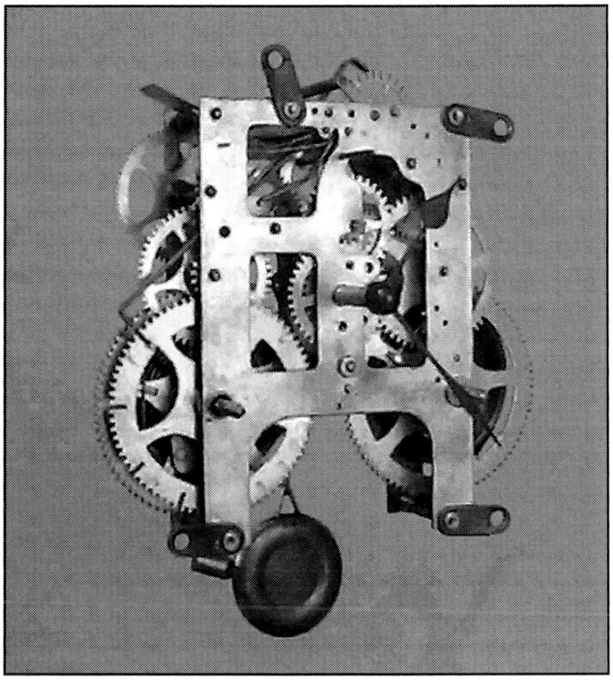

Fig. 19. This Gilbert movement has a passing strike on a bell at the half hour and a gong strike on the hour.

Figure 20, on the next page, is a rear view of the movement, showing the bell which screws onto a post on the rear plate. For illustrational purposes, most of the strike parts have been left out, except the strike hammers and the bell. The two hammer arbors cause repairers much difficulty on reassembly. They are placed close together, almost intertwining, with the brass wire hammer springs arranged in a special way.

A feature of this strike is that the cam rotates once per hour and has only one arm on it. This reminds us of an hour striker, but this clock *does* sound the half hours. This is possible because two strike lifting levers, one for the hour and the other for the half hour, are oriented 180 degrees apart. This allows the cam to lift a lever each half hour. The arrangement differs from one in which the train runs at the half hour; in that case, a cam with two arms lifts the lever each 30 minutes. Our present example has a cam with one arm that lifts two different levers, one each 30 minutes. The half hour hammer tail is simply lifted and then dropped at the half hour to sound the bell. Opposite this part is the lifting lever, which starts the hour run of the strike train.

Unfortunately, the assembly of the two strike hammers is tricky, and it makes the movement generally difficult to reassemble. One does not see this half hour bell striker every day, so it is hard to remember how the parts fit together. The wire springs, which provide hammer tension, must be hooked up right away, before assembly is started. (It's too difficult to hook them later on.) That makes the arbors tend to pop out of the holes during assembly.

Figure 21 shows the way in which the two hammer arbors fit together. They really do intertwine in a way that makes them a frustrating exercise to assemble. The wire tension springs make the job

Fig. 20. Rear view, showing half hour bell and hammer. Hour hammer is shown at the bottom of the photo.

even harder. Note that the spring for the hour hammer arbor loops over the movement pillar. The spring for the half hour hammer, however, loops over the hour hammer arbor. Check the springs and replace them if they appear likely to fail. It is probably not possible to install a new tension spring on a fully assembled movement, so take the opportunity to replace the springs, if necessary, while the movement is apart. This strike mechanism can be frustrating! Be patient, for the assembly may take several tries before everything fits.

The hour and half hour striking points are adjusted separately, after the movement is assembled. To check them, install the minute hand and allow the movement to run to the hour or half hour. Turning the hand yourself will make any test false, because of backlash in the motion work gears. Bend the hour lifting lever or the half hour hammer tail, depending on which needs adjustment, to achieve an earlier or later start to the strike. Never bend the minute hand to make an adjustment of the striking point.

FIG. 21. DIAGRAM OF HAMMER ARBORS, REAR VIEW, GILBERT STRIKE MOVEMENT

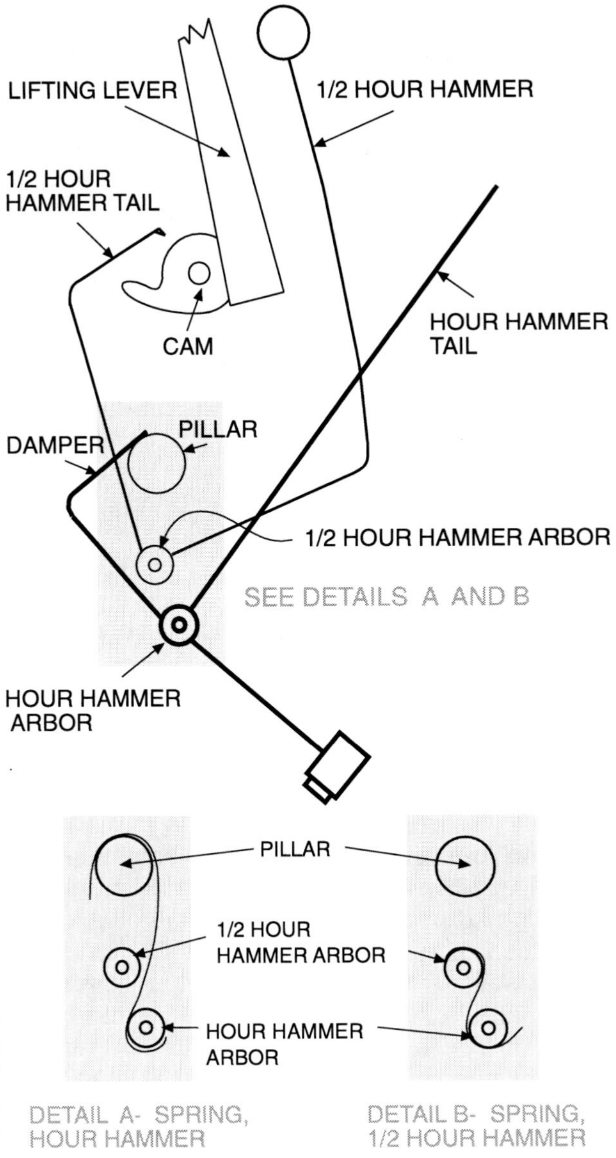

Bim-Bam Strike on Chime Rods

Many mantel and wall clocks made in the early 20th century feature a bim-bam strike. Gilbert's version is typical. Two chime rods, with one note higher in pitch than the other, are installed in the case instead of a single coiled wire gong. The tambour case is ideal for chime rods, since it provides the width required for the long rods. Other case styles allow for diagonally or vertically mounted rods.

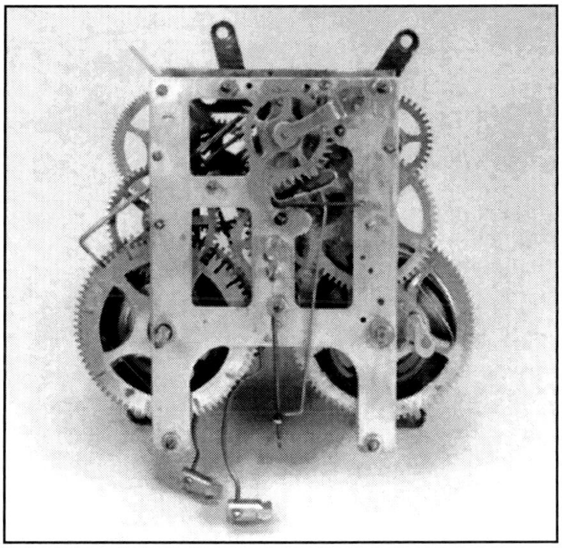

Fig. 22. Gilbert bim-bam movement

Figure 22 is a Gilbert bim-bam movement with the two hammers mounted for vertical chime rods. This style fits in a wall clock case.

In the bim-bam strike arrangement, each hour strike count consists of two separate notes: a strike on each rod. This two-note system is called "bim-bam". One o'clock is sounded as the high note followed by the low note. Twelve o'clock is sounded as 12 pairs of notes: high-low, high-low, and so on. The higher counts take a considerable time to complete!

The half hours are sounded as one strike on the lower note. This is a passing strike that does not involve the running of the strike gears to lift the hammer. The strike gear train is basically an hour striker like our previous example.

A separate half hour cam (A in Figure 23) is staked lightly onto the elongated back end of the center arbor. This cam can be twisted around to adjust the exact moment the half hour would be struck. Only one hammer has the long, curved tail which is lifted by the cam at the half hour.

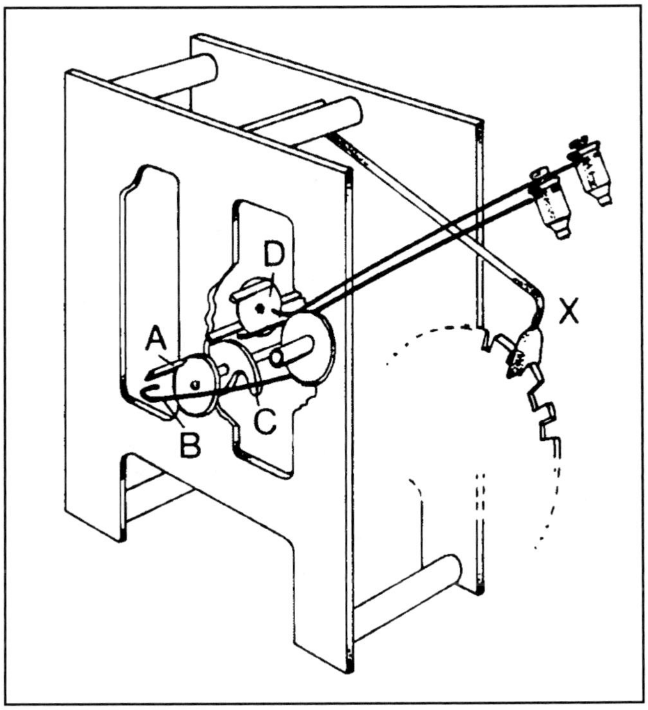

Fig. 23. Gilbert bim-bam hammer arrangement: ***A*** *half hour cam;* ***B*** *half hour hammer tail;* ***C*** *center cam;* ***D*** *hammer-lifting cam;* ***X*** *count lever.*

Listen carefully to the sound as you adjust the strike, for the hammers tend to bounce too much and cause a vibration in the tone. The half hour note may sound loud enough when the minute hand is turned through the half hour, yet be too quiet when the clock runs on its own. The half hour hammer tail (B in Figure 23) may be bent to adjust the amount of lift the hammer receives at the half hour. Bend the hammer shafts (the brass rods which carry the hammers) to increase or decrease the at-rest clearance between the hammer and the chime rod. This adjustment will determine whether the hammer will bounce too much or make a clean, single tone.

5

SETH THOMAS:

NO. 89 STRIKE
O.G. WITH STRIKE AND ALARM

The Seth Thomas No. 89 movement exists in a number of versions with different pendulum lengths and hammer arrangements. Some are stamped with the "89" followed by a suffix indicating a particular model. A modified No. 89 movement is applied in an interesting way in the Sonora Chime; the No. 89AM movement functions as the time and hour strike portion of that famous series of clocks. Other versions of the movement, some with a solid back plate, will cross your bench sooner or later. Whatever the modification, the basic movement remains the same. The No. 89 is a movement you will encounter often in various types of cases, plain and fancy.

Figure 24 is a photo of a basic No. 89. The number is stamped on the lower rear part of the back plate, and also directly in front of it on the lower inside surface of the front movement plate. It features recoil pallets fitted between the clock plates. Mounting brackets riveted to the front movement pillars are typical for a mantel clock installation, with the movement screwed to the inside of the wooden dial support.

For details of the strike train parts, refer to Figure 25 as we go along. The count wheel (1) is mounted directly in front of the strike main wheel (2). As the clock strikes, the drive gear (3), mounted on the second arbor, rotates the count wheel. To understand the strike levers, locate the two arbors which support three levers each. Here we run into a similarity with the Ingraham movements which have the same arrangement. One arbor has the lifting lever (15), unlocking lever (11), and the warning lever (16). The other arbor has the drop lever (10), count lever (4), and the locking lever (9). Here the similarity ends, for each strike movement seems to be a combination of features found in many others. Once you have seen most of the ways to arrange levers, warning and locking features, and the count wheel, it is just a matter of taking note of how a given movement is designed before you take it apart. The No. 89 movement has a cam (5), as we would

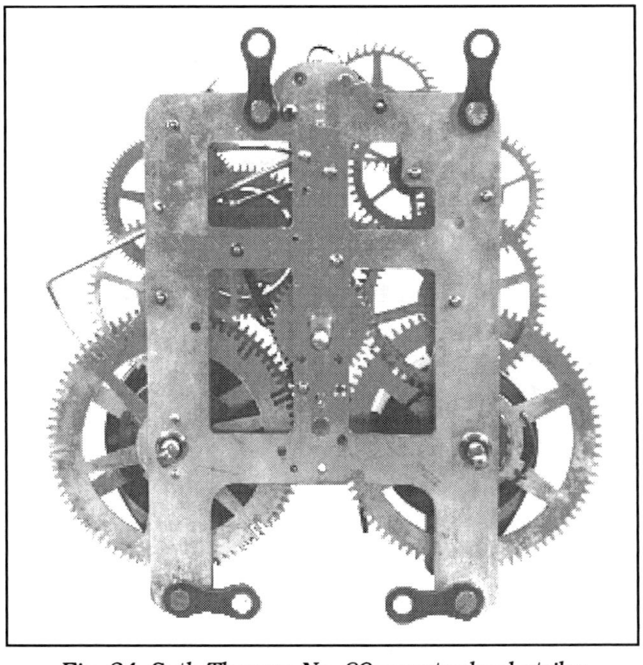

Fig. 24. Seth Thomas No. 89 count wheel strike

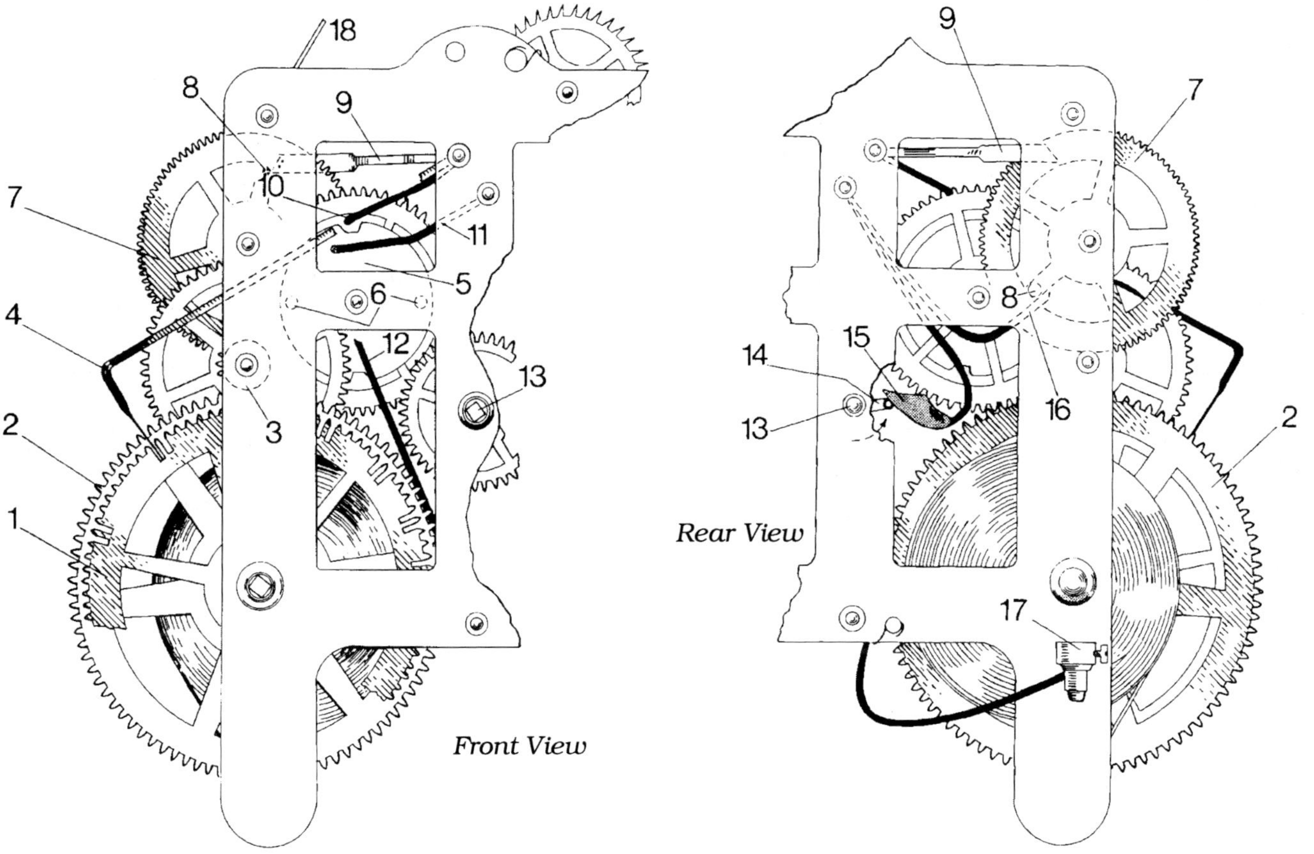

FIG. 25. SETH THOMAS NO. 89 STRIKE MECHANISM

1 COUNT WHEEL
2 STRIKE MAIN WHEEL
3 COUNT WHEEL DRIVE GEAR
4 COUNT LEVER
5 CAM
6 HAMMER LIFT PINS
7 LOCKING WHEEL
8 LOCK PIN
9 LOCKING LEVER
10 DROP LEVER
11 UNLOCKING LEVER
12 HAMMER TAIL
13 CENTER ARBOR
14 CENTER CAM
15 LIFTING LEVER
16 WARNING LEVER
17 HAMMER
18 FLY

expect, with two slots. The two hammer lift pins (6) are fastened to the cam, raising the hammer tail (12) as the strike train runs. The locking wheel (7) has one pin on it, the lock pin (8) which serves both warning and locking purposes.

Now that we have taken a survey of the strike train design, we can proceed with the repair. It is wise to check for worn pivot holes first, before you do any disassembly work. The No. 89 is especially prone to excessive wear on the front pivot hole for the third strike arbor. This is the arbor which carries the cam (5). The escape arbor pivot holes are also very likely to be worn. In fact, wear on these holes seems to be what stops many of the No. 89 movements after so many years of service.

Before going any further, let down the mainsprings into clamps. Make sure the springs are *completely* let down. If the springs are dirty and sticky, they can fool you. Next, look for the small spring studs on the movement plates. There are three of them, one each on the front and back near the escape arbor (for the strike levers), and one on the lower rear, for the hammer arbor. The brass wire springs are hooked onto these convenient studs, making it unnecessary to wrap the wires around pillars. Unhooking the wires now will lessen the chance you will distort or break them as you separate the plates. Replace any questionable wire spring with a new one.

Separate the plates and remove all the parts. Do not forget to check the clicks—many are found to be loose on the No. 89 movements. The click springs should also be checked. Clean the movement thoroughly and polish any rough pivots. Install bushings as needed. Finally, after a considerable amount of work, you will be ready to put the movement together again and adjust it. Assemble the movement, but do not wind the mainsprings. Check the strike train in the locked position. With the count lever (4) in one of the deep slots in the count wheel, the drop lever (10) should be centered in one of the slots in the cam (5). The lock pin (8) should be at the locking lever (9) as shown in the Figure 25 front view. Separate the plates and ease the locking wheel out of place to correct any misalignment you find.

Once this adjustment is correct, the strike train should work. There may be a number of other adjustments required, however. Check the pillar nuts for tightness, and then wind the mainsprings partially. Install the minute hand and turn it slowly clockwise to the warning position. Warning should occur five to seven minutes before the hour and half hour. Bend the unlocking lever (11) to give earlier or later lift, to obtain a warning in this range. Check the warning lever (16) to make sure the lock pin catches it safely. The Figure 25 rear view shows the placement of the pin and lever at warning.

Turn the hand to cause striking to begin. Observe whether the count lever dips neatly into the center of each slot, especially the deep slot at the end of the strike sequence. Bend the end of the count lever carefully to obtain the right action. As a point of reference, the count lever should point at the center of the count wheel. That is the only way the lever can stay centered in the slots, which are of course radial. If the count lever does not safely clear the tips of the count wheel teeth during striking, bend the drop lever (10) downward slightly to correct the problem.

One of the most important adjustments is that of the locking lever (9). You may have to bend it up for less locking action if the pin glances off the lever as the clock strikes. Bend the lever down if the locking action is not safe. Needless to say, if the strike train does not lock properly it will run on occasionally (or continuously) and the strike count will be wrong. The locking action must be closely observed.

Another adjustment which is often overlooked is that of calibrating the minute hand so the clock will strike as exactly as possible on the hour and half hour. Some clocks have minute hands with moveable bushings which permit adjustment of the hand itself. The Seth Thomas movement, like most American clocks, does not have this feature. Bending the hand will distort or break it, so the only way to make the adjustment is to go to the lifting lever (15). First, install the minute hand and pin. Allow the movement to tick (rapidly, with the pendulum off) to the strike point. Turning the hand will not give the same result, because of normal slack in the motion work. Just observe the exact position of the hand when the center arbor (13) turns just enough to cause the center cam (14) to drop the lifting lever. If the striking begins early, bend the lifting lever down. The bending point is not at the end of the lever, but about 1/3 of the way up from the hooked end. If striking is late, bend the opposite way.

The No. 89 movements are well made, although there are several parts which frequently wear out. The front pivot hole of the third strike arbor has already been mentioned as a place that is often worn. Most No. 89 movements will require pivot polishing and a bushing to be installed at this location. The escape pivot holes and those of the two arbors below it are also wear points. The escape pivot holes, in addition to worn pallet faces, probably stop most of the No. 89 movements that are brought to the shop for repair. Otherwise, they are durable movements. An especially helpful feature is the use of brass posts on the front and rear plates. The brass wire springs which provide tension for the strike arbors are wrapped around these posts. It is unfortunate that all movements did not have these posts installed at the factory. They make it possible to anchor the wire springs neatly and without danger of getting them caught in the wheels. If care is used, the wire springs can be unhooked from the posts, then later hooked back on without kinking or damaging the springs.

The clock is almost ready for a test run. Remove the mainspring clamps and complete the oiling of the movement. Check the wire springs once again to be sure they are secure. Install the movement in the case, set the pendulum beat, and adjust the hammer for correct tone with the gong. Wind the mainsprings fully and test for a full week. Any strike counting errors indicate you have some more adjustments to make. On this movement, an intermittent counting error is likely to be caused by either the count lever not centering in the slots, or the locking lever not engaging the pin deeply enough.

Seth Thomas OG Strike with Alarm

Fig. 26. The Seth Thomas 30-hour OG clock

The OG clock is distinctively American in character—strong, direct, and functional. The Seth Thomas OG in Figure 26 is a weight driven 30-hour clock which represents the type very well. One can imagine the clock's owner in an earlier time, inserting the crank key every night in ritual fashion, never forgetting. Although I liked the 30-hour OG I once owned, I would inevitably fail to raise the weights after a few days of regular winding. When I would remember, I always had to take the time to get the strike to agree with the hands again, and to reset the time. Finally I stopped using the clock entirely!

Beginning clock repairers are attracted to this style of clock because of the large gears, overall simplicity, and the absence of mainsprings. Indeed, the weight driven OG is a good place to start. It offers a count wheel strike train, a strip pallet recoil escapement, and even an alarm mechanism in the present example.

The movement rests on a seatboard which is slotted into the vertical wooden supports you can see in Figure 27. The weight cords go up to the top of the case, over wooden pulleys, and down to the weights. During the course of a winding period (30 hours or eight days, depending on the clock), the weights move all the way from the top of the case to nearly the bottom. There are only two train wheels plus the fly in the 30-hour strike train, as shown in Figure 28. The main wheel (4) serves as the pin wheel. The second wheel is the locking wheel (7), and the drop lever (12) locks on the cam (13). The warning pin (15) is fastened to the fly arbor instead of a wheel. Eight-day OG clock movements look very much the same as the 30-hour variety, except for the additional wheel in each train.

To begin the repair, first remove the hands and dial. Take off the weights and pendulum bob, and don't forget to unhook the control wire for the alarm mechanism. Look for the slotted wooden block fastened to the case just above the movement. Remove this block and then pull the movement and

Fig. 27. Movement shown mounted in the case

seatboard out of the clock together. After taking off the two seatboard screws and the seatboard, you are ready for disassembly. Examine the movement first, if you wish, to locate the pivot holes requiring a bushing. Make sure you accurately judge the direction of wear on each hole. This makes it easier to "draw the hole" back to its original center with a pointed file before you ream the hole for a bushing. Disassemble the movement and clean it thoroughly. The movement illustrated here had dirt packed in

FIG. 28. SETH THOMAS
30-HOUR OG MOVEMENT

1 CENTER ARBOR
2 ALARM DISK
3 ALARM TRIP LEVER
4 MAIN WHEEL (PIN WHEEL)
5 COUNT WHEEL
6 COUNT LEVER
7 LOCKING WHEEL
8 FLY
9 LIFTING LEVER
10 CENTER CAM
11 UNLOCKING LEVER
12 DROP LEVER
13 CAM
14 WARNING LEVER
15 WARNING PIN
16 HAMMER
17 STABILIZER
18 HAMMER TAIL
19 HAMMER SPRING
20 HAMMER LIFT PINS (13 PINS)

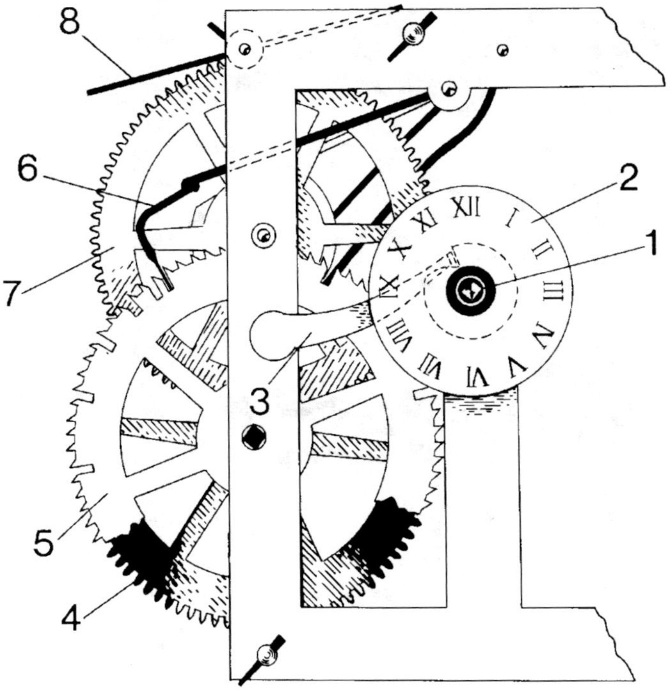

Front View

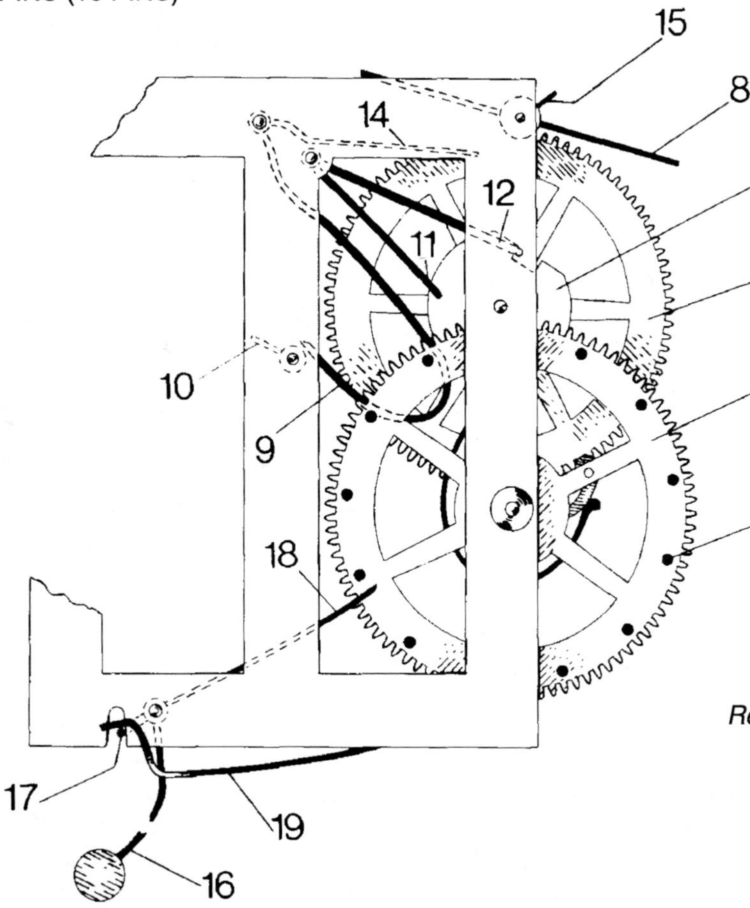

Rear View

the lantern pinions, but fortunately there was very little rust. One click rivet was loose and had to be tightened. The pivots required only slight polishing, as did the pallet faces. Bushings were needed for the time and strike second arbors, front and rear, and the escape arbor pivot holes. The weight cords needed to be replaced.

The alarm mechanism consists of a small main wheel with a mainspring, geared to a pinion. On the same arbor with the pinion is a wheel with pointed teeth, and above that is a strip pallet. The winding setup for the alarm is fitted with a stop-works which permits a full travel of about 2-3/4 turns. This means it takes only that number of turns to wind the alarm, and the raucous alarm is limited, mercifully, to about a ten-second run. More about the alarm after we complete the movement overhaul.

The movement looks simple enough to assemble, but it can fool you. Figure 28 shows how the two sets of strike levers intertwine. Observe that the drop lever goes in the slot in the cam. The lifting lever (9) must be in line to meet the center cam (10), and the lifting lever must be able to raise the unlocking lever (11). The warning lever (14) goes above all the other levers, so it can stop the warning pin (15) at the right time. It is a good idea to wind a brass wire spring onto the arbor which carries the count, drop, and unlocking levers. Slight spring pressure will assure positive locking action. Wind the spring onto the arbor while the clock is still apart, but do not hook the spring over the movement pillar until the assembly is complete. An arbor with spring pressure on it will continually pop out of the holes all the while you are trying to put the movement together.

Do your best to assemble the movement with the strike train in the locked position, but don't worry too much about it. You can always make adjustments later. Install all the arbors into the rear plate, except the escape wheel, which should be in the front plate. With this Seth Thomas, you must make sure the alarm trip lever (3) is above the center arbor (1). If you leave the trip lever below, you cannot move it up over the center arbor after the plates are together. The hammer spring need not be hooked up until after the assembly is complete.

With the clock movement now all back together again, it is time to see how close you are to having the strike train correctly set up. Using finger pressure, turn the strike train gears. First, notice how the count wheel (5) is moved along in steps, as the single pin near the center of the locking wheel moves it ahead with each revolution of the locking wheel. The pin moves the count wheel just far enough each time, and the wheel stays put until it is moved again before the next hammer stroke. This means we do not have to be concerned over the synchronization of the count wheel's movements versus the dips made by the count lever. It is automatic.

Next, watch the hammer (16) rise and fall. You should hook up the hammer spring (19) if you haven't already done so. When the strike train locks at the end of a sequence, the hammer tail (18) should be free of the hammer lift pins on the main wheel. If the hammer ends up "on the rise", then separate the plates enough to allow yourself to change the mesh of the main wheel and the pinion above it. This changes the relationship of the cam, which determines the locking point, and the main wheel. You may find, however, that even though the hammer tail does clear the pin at the end of striking, it still comes to rest against the next pin. Adjust this by bending the stabilizer (17) slightly. It is this part which determines the rest position of the hammer by rubbing on the cutout portion of the rear plate. In addition, the stabilizer prevents excessive hammer bounce.

Now turn your attention to the warning. The fly arbor carries the warning pin (15). You may have to change the mesh of the fly pinion with the locking wheel to establish a correct warning run. When the gear train is locked, the warning pin should be at an 11 or 12 o'clock position so there will be a run of the gears at warning. The warning occurs as the lifting lever (9) pushes up on the unlocking lever (11), moving the drop lever (12) out of the cam slot.

Do not neglect to install the minute hand so you can check the strike point. Striking should begin as nearly as possible on the exact hour. There is no hand bushing to adjust, so you will have to bend the lifting lever. Looking at the lever from the back of the movement, bend it at a point below where it contacts the unlocking lever. Bend down to make the strike go off later and up to make it go earlier.

If someone has been into your movement before, bending all the levers, you will have some correcting to do. The count lever should enter the count wheel parallel to the sides of the slot, with the end of the lever pointing at the center of the wheel. The drop lever should lock about one full diameter down in the cam slot. Check the adjustment by running the gear train slowly. As the clock counts hammer blows, the count lever enters the shallow slots between count wheel teeth, allowing the drop lever to dip only slightly into the cam slot. If the drop lever grazes the corner of the slot, the locking is too deep; bend the drop lever up to correct it. Bend the lever down only if it doesn't lock safely in the cam slot as striking ends. The strike warning should occur about five minutes before the hour. Bend the unlocking lever up to make the warning come later, down to make it come earlier. The warning pin should not cause much trouble as long as it locks on the warning lever when it should and clears it safely during striking.

Finish up by adjusting the alarm mechanism. The alarm trip lever rides on the cam behind the alarm disc (2). The disc is marked with the twelve hours as shown in the Figure 28 front view. To set the alarm, turn the disc until the desired wake-up hour lines up under the hour hand. The long control wire from the trip lever to the alarm mechanism keeps the pallets on the alarm from operating, until the trip lever drops into the slot behind the alarm disc. Then the control wire becomes slack, allowing the alarm to run as long as the spring can power it. Too much slack means the alarm will start right up whenever it is wound; with too little freedom, the alarm will not release at all. Bend the wire on the alarm mechanism until it operates correctly.

During the test run, check for an incorrect strike count or an extremely uneven fall between the two weights. These are indications of an adjustment problem in the strike mechanism. You don't have to test the clock for a long period of time, since the full winding cycle is only one day. Just run the clock for a sufficient number of days to convince yourself there are no problems—that is, if you can remember to wind it every day!

6

ANSONIA

As most collectors are well aware, Ansonia Clock Co. made a great variety of clocks. Repairers know Ansonia for the Brocot "visible" escapement used in so many of the movements. The strike mechanisms were mostly count wheel types.

Our subject for this chapter is an Ansonia china-cased clock from the 1890's, shown in Figure 29. Although many such cases are decorated in colors such as hot pinks and lurid greens, this particular one has more muted colors: purple and yellow pansies are painted on an off-white background with green trim. The Brocot escapement can be seen in the dial, just below the numeral 12.

The front of the movement is shown in Figure 30, giving a closer view of the Brocot escapement. It is fascinating to watch the pallets swing back and forth, engaging and releasing the escape wheel teeth. Note the arched shape of the movement plates, typical of Ansonia.

Figure 31, the rear view of the movement, shows the small cutouts on the rear movement plate. It is still difficult to see and adjust the parts of the strike mechanism which are mostly hidden. The plate bears stamped patent dates of June 14, 1881 and June 18, 1882 and the number 4-1/2.

Figure 32, including Details A, B and C, shows the parts of the strike mechanism. The count wheel has 90 teeth and counts the hours and half hours. This means the strike train runs on the half hours as well as the hours; the half is not simply a passing strike in which a hammer tail is picked up and dropped to sound the gong once. A small brass gear on the second arbor drives the count wheel which is mounted concentric with the winding arbor.

Strike locking occurs as the locking lever, shown in Detail A, stops the lock pin. The same arbor carries the drop lever, unlocking lever, and count lever. The warning lever (see Detail B) catches the pin to temporarily stop the gears, waiting for the hour or half hour release point.

Fig. 29. Ansonia La Normandie china-cased clock

Detail C shows the hammer, hammer tail, and damper. The damper is a piece of stiff wire made from the same steel from which most of the levers in the clock are made. As the clock strikes, the damper rubs against a slot in the front plate to limit the bouncing of the hammer.

The overall functioning of the strike train is similar to other count wheel types. The cam has two slots for the drop lever and two hammer lifting pins. A center cam with two pins starts up the strike mechanism each half hour.

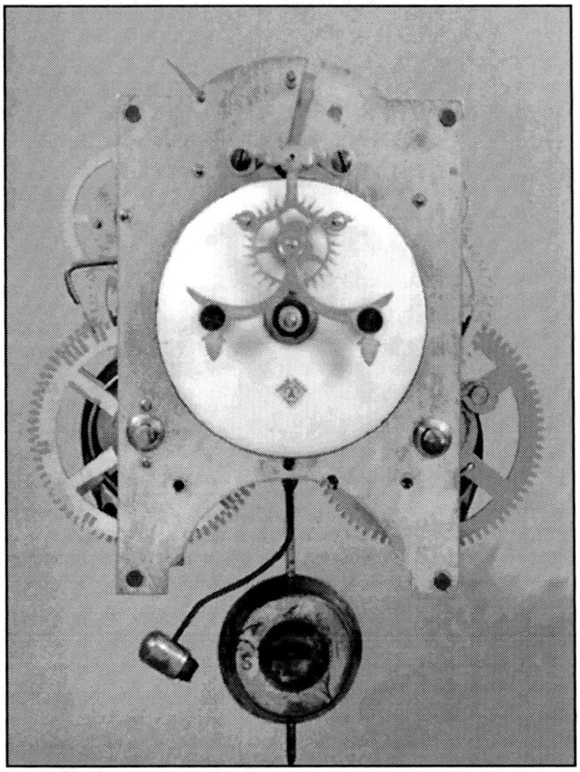

Fig. 30. Front view, Ansonia strike movement

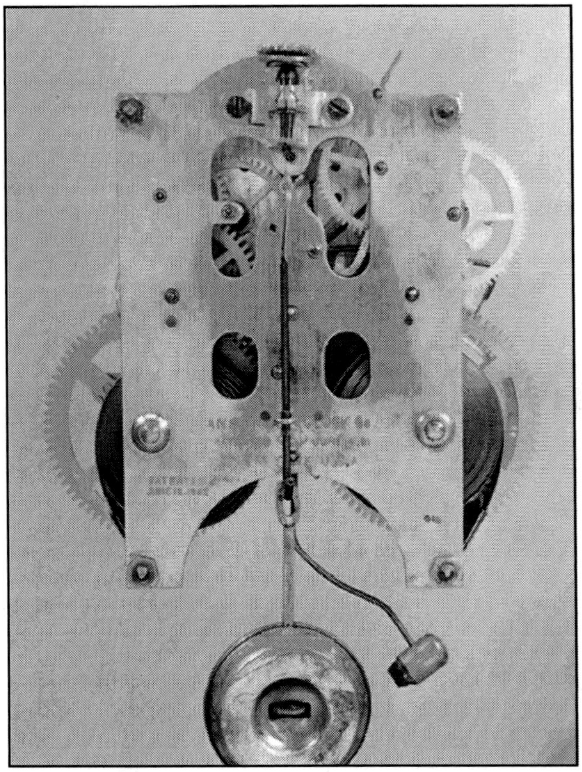

Fig. 31. Rear view of the movement

Lifting/Warning Lever (Detail B, Figure 32)

The part of the Ansonia strike mechanism most likely to cause repair troubles is the stamped brass piece shown in Detail B, Figure 32. Most Ansonia movements of this style have a similar piece, although some are shaped differently, like Figure 33.

The lower portion is the lifting lever with the warning lever at the upper left. The entire piece is fastened to a steel post on the inside of the back plate. The drawings show a front view as the part is seen when the front plate is removed.

On most of the Ansonia movements, the lifting/warning lever operates only by gravity. As this piece is lifted towards the warning point, the lock pin comes to rest against the warning lever. At the striking point, the brass lever drops, allowing the strike to begin.

Since the post is riveted over to fasten the brass lever in place, it is not practical to remove the lever for cleaning. It is simply placed in the cleaning solution as part of the rear plate. The cleaning solution and rinse must completely degrease the parts so the lever will operate from its own weight. If it becomes sticky from incomplete cleaning, the movement will occasionally (or always) fail to come off the warning. Recheck the operation of the lever several hours after cleaning and drying (better yet the next day) to make sure that grease or dirt remaining on the post has not set up like glue.

Rather than depend on gravity alone, it is better to have a wire spring on the lifting/warning lever. Figure 32 shows the way a wire spring seems originally to have been set up on some of these movements. Here's how a spring can be made. With the movement disassembled, wrap one end of a piece of 30 gauge brass spring wire around the Detail A arbor, which carries the count, drop, and locking levers. Make several counterclockwise turns of the wire and leave enough wire for hooking at the other end. (The Appendix gives more information on making brass wire tension springs.) After the movement has been assembled, attach the free end of the wire spring to the "hook for wire spring" shown in Detail A. The result is a single spring that operates both the Detail A and Detail B levers.

It is a good idea to add a wire spring to the lifting/warning lever even if the movement appears never to have been fitted with one. Figure 33 is an example, an Ansonia lifting warning lever with no hook on it. (It performs the same functions as Detail B, Figure 32.) In Figure 33, a spring can be added. Loop the end of the wire over the lever at the

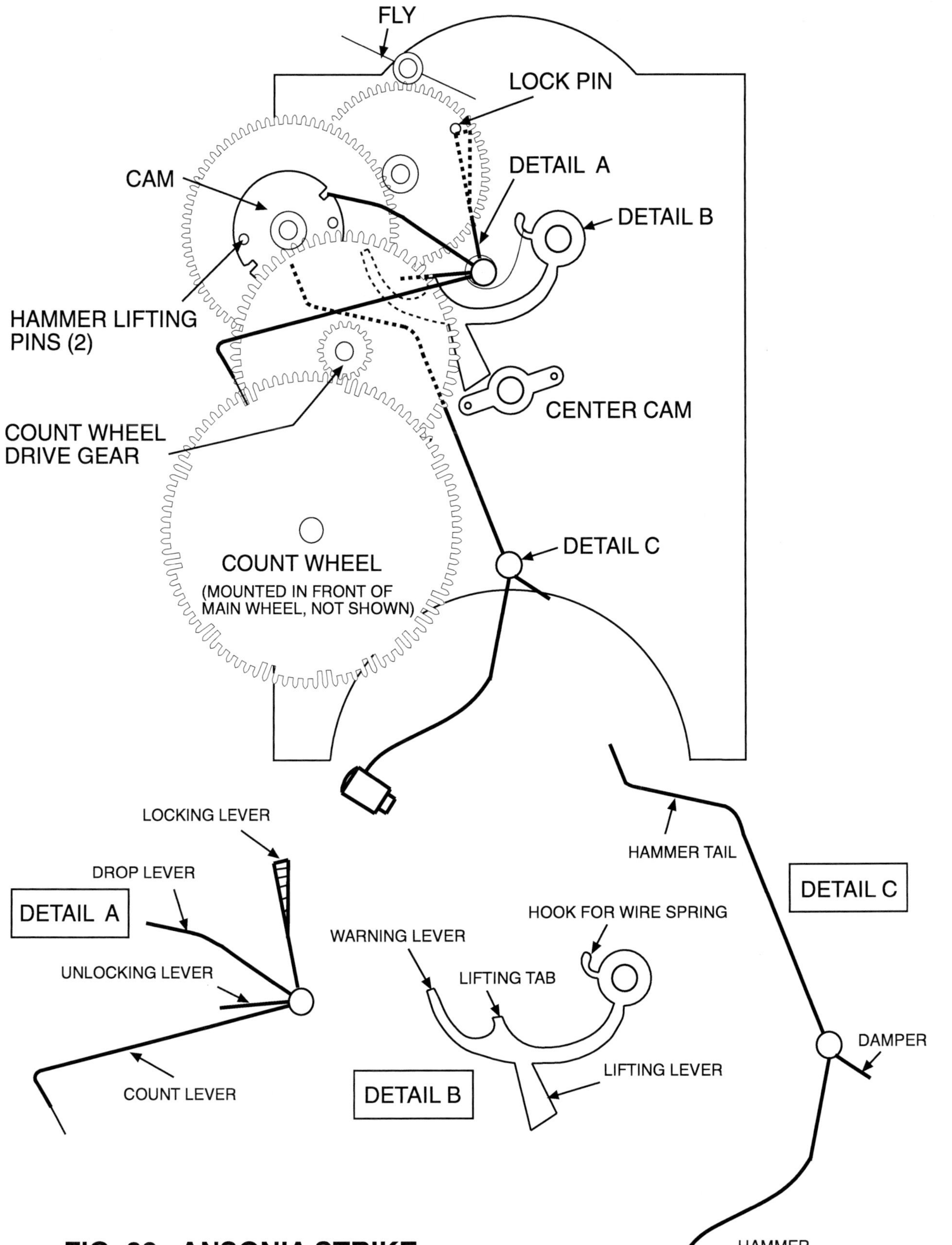

FIG. 32. ANSONIA STRIKE MECHANISM, FRONT VIEW

sharp angle to the left of the post. Wrap it counter-clockwise around the post and then fasten to the strike lever arbor.

The wire spring may seem to be a small point, but a lever which becomes sluggish after a few weeks will fail to work by its own weight alone. That means the clock must be taken in again and repaired. Remember that the wire spring can only be fitted when the movement is apart, so you should add it then if you have any doubts about the free motion of the lever. If the lever has a hook indicating that a spring belongs on it, be *sure* to install one.

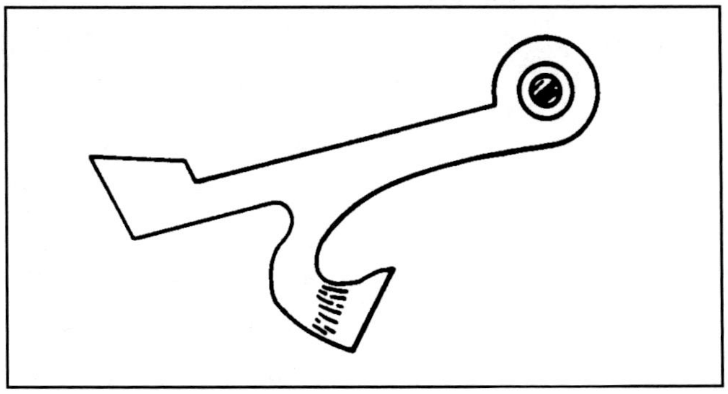

Fig. 33. Variation of the lifting/warning lever from Figure 32, Detail B. This one has no hook for a wire spring.

7

SESSIONS

After the Sessions Clock Company began carrying on the work of E.N. Welch in 1903, it manufactured a large number of clocks. Black mantel clocks and the rectangular "store regulator" clocks were especially common. The striking movement pictured in this chapter is so plentiful that almost all repairers will have encountered it.

The Sessions design is so similar to the Gilbert (Chapter 4) and the Seth Thomas No. 89 (Chapter 5) that there is no need to draw and label each part in the present chapter. Sessions has been included here because the display model I made for the video *Repairing Eight Day American Striking Clocks* provides a unique working view of the American striking movement.

Figures 34 and 35 are views of this model, which has a clear plastic front plate invisible to the video and still cameras. The mechanism runs and strikes, providing an unobstructed view of the parts. One does not realize how difficult it can be to see the parts to be adjusted—until the front plate disappears!

Early in the video project, the producers and I decided that the various parts of an actual movement could best be shown in a "see-through" model. A scrap Sessions movement was disassembled and cleaned of all rust, dirt, and oil. One of the clock plates was used as a template for drilling holes for the front pivots. Brass bushings were pressed into the clear plate to act as bearings for the front pivots. The original steel pillars were removed and a new set made up. The two top pillars were plastic, for appearance; the center and two bottom pillars were brass, for strength. This description simplifies

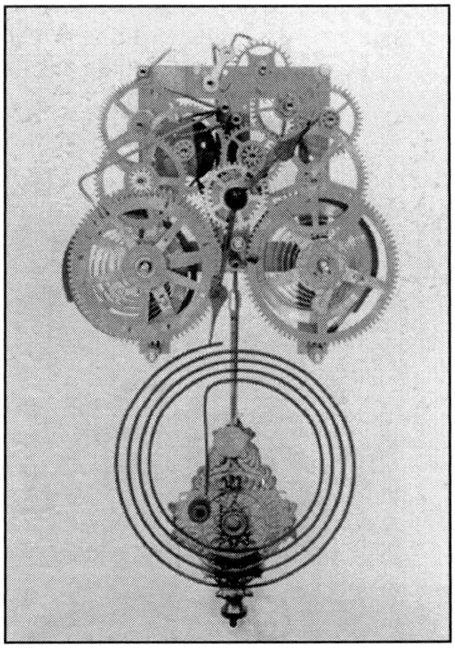

Fig. 34. Sessions movement with see-through plate

the actual making of the model, which required many hours of work. The result was certainly valuable as a visual aid and learning device.

Anyone who would like to make this kind of see-through model should keep in mind a few suggestions. I have not made a study of the strength of the finished product, which was never intended for regular service as a household timekeeper. Remember the strength of the two mainsprings, especially as you design the pillars and the method by which they will be attached to the brass rear plate and the see-through front plate. The material used in the Sessions project is a G.E. product called "Margard", which may be similar to "Plexiglas". The sheet material is about 3/16" thick and was purchased from a plastics dealer. It is recommended that one of the clock plates be used as a drilling template for the holes in the plastic plate. Attempts to locate the new pivot holes by measurement or depthing may be inaccurate, as I found out early along. In any case, drilling with the aid of a template is a fast and workable method.

What the Model Shows

Several aspects of American count wheel movements are revealed by the see-through model. There is an overall impression of many wheels and levers being fitted into a space which is small, considering the size of the wheels. The two gear trains are not arranged straight up, but are arranged with the gears staggered to lessen the overall height of the movement. The plates are stamped out in such a way that there is only enough brass left in them to support the pivots, and no more. Visibility of parts may be an issue for repairers, but one suspects that conserving brass was the main goal of the makers.

Fig. 35. Closeup view of the Sessions strike train

Another aspect, revealed in the close-up view of the model (Figure 35) is the interlocking system of strike levers. It is important to see them for what they are, a group of soft steel wires which are easily bent from the correct positions. This contributes to the difficulty every repairer experiences with the levers; since the levers are easy to bend, people bend them. In contrast, flat stamped levers are not as readily distorted.

Sessions Clicks and Rivets

A special problem in the Sessions movement is the relatively weak click and rivet. It is perhaps unfair to criticize parts which can function for fifty years or more, but it seems that the Sessions setup is less sturdy than others. Each click and rivet must be carefully checked, for it is common for the rivet to become loose under the stress of the mainspring. At some point enough distortion occurs to allow the click, which is made of rather thin brass, to slide over the top of the ratchet wheel, causing a failure. A new click and rivet will be required.

In many cases the original click has developed burrs and worn spots from long years of wear. If the mainspring is removed from the wheel, the click can sometimes be cleaned up with a small file and made as good as new. Do not shorten the click or change the angle on the tip, which should fit the shape of the ratchet teeth as closely as possible. Do not invite further problems by installing a new mainspring of the common type measuring .018" thick by 3/4" wide by 96" long, which is too strong for the movement. Suppliers have 3/4" wide loop end springs which are .0005" to .002" thinner and either the same length or longer than the .018" by 96" spring. There is usually nothing wrong with cleaning and reusing the original mainsprings, provided they are not rusty.

8

HAMBURG-AMERICAN

We turn now to a familiar style of German wall clock, sometimes called a "box clock". Our subject was made by the Hamburg-American Clock Co. The case, shown in Figure 36, is 12 inches wide by 26 inches high and must have included a top piece. The movement (Figure 37) is screwed to a seatboard which slides into brackets on each side of the case. A gong is fastened to the back of the case. Refer to Figure 38 to identify the parts.

The movement is a count wheel type with the count wheel located between the clock plates, toward the rear. You may be more accustomed to seeing this part near the front, as in an American striking clock, or mounted on the back of the movement in the French style. The two mainsprings are enclosed in barrels.

Mainspring troubles seem to account for many of the problems in the older German clocks in this category. Eventually the springs break, and severe movement damage often results. Several teeth may be torn from the barrel, or the first lantern pinion is destroyed, or the second arbor is bent. If only the spring is broken, consider yourself fortunate! The movements are otherwise durable.

It is always a good idea to take a look at the design of a movement before you take it apart. A glance at the Hamburg-American movement reveals the count wheel, as I have noted. Beyond that, the method of strike locking is especially important. There is no locking pin to stop the gear train. Instead, the drop lever doubles as a locking lever by seating itself against the cam. We saw this setup in the New Haven movement featured in Chapter 3.

Fig. 36. Hamburg-American wall clock

Repair and Reassembly

Begin the repair by letting down both mainsprings. The mainspring barrels cannot be removed without taking the clock completely apart. Remove the mainsprings and check them for cracked ends or the distorted, "helixed" appearance which signals replacement. Carefully check all barrel teeth and the pinions, arbors, and wheels above them. Inspect the click wheels for damaged teeth. Clean the movement and install the mainsprings in the barrels. After any bushing and pivot work, you are ready for reassembly.

Fig. 37. A view of the count wheel movement

To put the clock back together, install the parts into the rear plate. The two barrels go in first, followed by the time train second wheel and then the center arbor. The time train is placed low, leaving most of the upper half and one side of the movement for the strike train. This open design makes the movement a pleasure to handle. Continue adding the other parts, referring to Figure 38 for the strike parts as you go along. Leave out the pallets and back cock from the time train; these go in last. Do not forget to put in the hammer arbor (17) which is "between the plates". It protrudes out the back of the movement to receive the hammer. If you leave the hammer arbor out, you will have to take the movement apart again to put it in. The same is true for the arbor which carries the drop lever (12), the unlocking lever (13) and the count lever (14). When all the parts are in place, add the front plate and fasten the pillar nuts finger tight. Next, the lifting lever (4) goes on the front of the movement. Push the cannon pinion (3) onto the center arbor (2). Note that the hand tension is provided by the "crimp" in the brass sleeve. You should feel slight resistance as you push it on, and you can readily check the hand tension by installing the minute hand and trying it. The minute wheel and hour wheel can be installed now or left for later. (They are omitted from the Figure 38 front view to show the underlying parts more clearly.)

Strike Adjustments

With the assembly of the movement now complete, it is time to adjust the strike train. The strike train wheels must be set up correctly with respect to each other. The second wheel is on the same arbor with the count wheel; the third wheel has the hammer lifting star (16) mounted behind it; the locking wheel (8) shares its arbor with the cam (7); and the warning wheel (11) has the warning pin (10) mounted on it. If the movement is to count the hours correctly, the relationships must be right.

Begin by turning the strike train by hand. Watch the hammer rise and fall. There is no need to install the hammer shaft and hammer head at this time. If you put the wire spring in place, the hammer arbor will snap back as the hammer lifting star releases the hammer tail. The count lever touches the count wheel at the top. In the drawing, Figure 38 (bottom), the end of the count lever can be seen through the small hole in the movement plate. The slots in the wheel are the locking points at the end of each strike sequence. The width of the slots is a tip-off that the half hours are counted on the wheel, rather than by a "passing strike" arrangement.

When the strike train stops at the end of a sequence, observe the hammer tail. It should be about midway between points on the hammer lifting star. If the hammer tail is still in contact with one of the points, the hammer will be left in the raised position. To change the hammer position, separate the plates enough to move the locking arbor pivot out of its hole. Double check that the mainsprings are still fully let down before you begin. The locking point is determined by the cam, and your object is to disengage the cam while you move the hammer lifting star, until it releases the hammer tail. Put the plates back together and try the strike action again. Continue with this procedure until the hammer tail is free of the hammer lifting star at the end of each strike sequence. Check to be sure the count lever goes into the slot in the count wheel and counts the half hour in the same slot. This adjustment is not an especially fine one, and it usually takes care of itself as you adjust the hammer lifting star and the cam.

Next look at the warning wheel. As the strike sequence ends, the warning pin should come to rest at a 10 to 12 o'clock position, when viewed from the front. This will allow approximately a half-turn warning run. Figure 38 (top) shows the pin in this orientation. To adjust the pin, separate the plates slightly and ease the front pivot of the warning wheel out of its hole. Keeping the drop lever locked in the cam slot, move the warning wheel to the correct

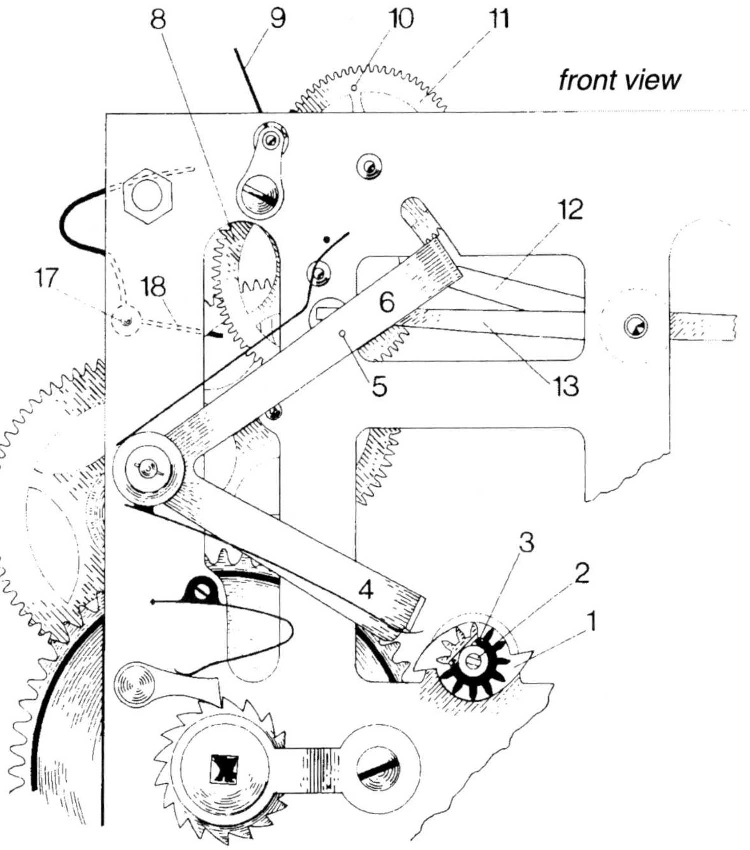

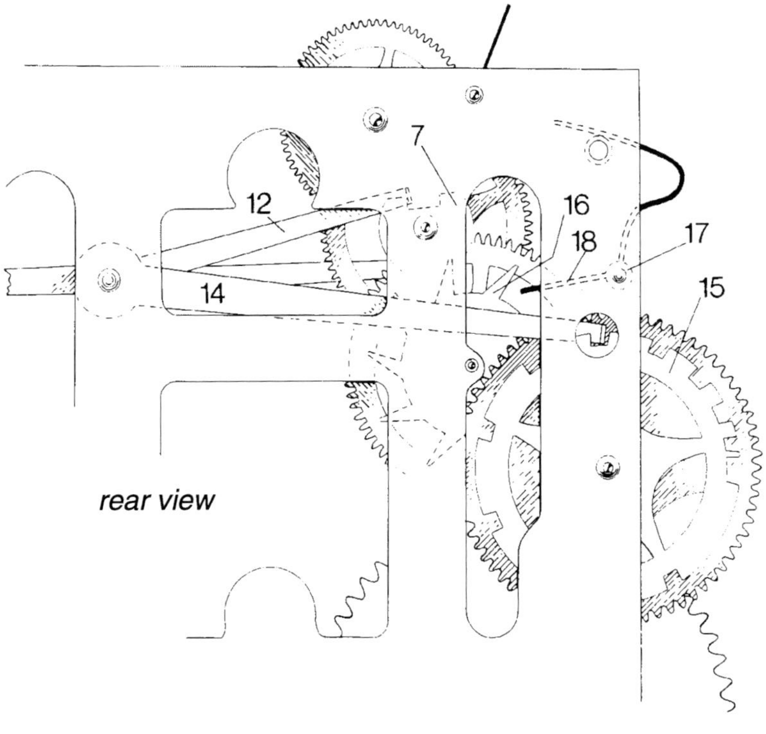

FIG. 38. HAMBURG-AMERICAN STRIKE MECHANISM

1 STAR CAM
2 CENTER ARBOR
3 CANNON PINION
4 LIFTING LEVER
5 LIFT PIN
6 WARNING LEVER
7 CAM
8 LOCKING WHEEL
9 FLY
10 WARNING PIN
11 WARNING WHEEL
12 DROP LEVER
13 UNLOCKING LEVER
14 COUNT LEVER
15 COUNT WHEEL
16 HAMMER LIFTING STAR
17 HAMMER ARBOR
18 HAMMER TAIL

position. Put the plates back together again and check the adjustment.

At this point, you should not have to separate the plates again. Tighten the pillar nuts now, before it is forgotten. Install the click wheels and their covers, the minute and hour wheels, and the hammer shaft. Add the pallets and back cock, and adjust the pallet depth carefully. Wind the mainsprings partially, so you can continue to adjust the movement. Oiling can be done at this time.

As with any striking clock, the strike points should be checked to assure that the clock will strike as exactly as possible on the hour and half hour, rather than early or late. Most of the German clocks have movable minute hand bushings which you can use to make the adjustment. It is often difficult to move the minute hand bushing with pliers; it is better to file a square on the end of a piece of brass rod which you can hold in the vise. This gives you the leverage you need to move the bushing without damaging it. Make sure the bushing is tight enough when you are done adjusting it. If the hand slips on the bushing as the customer sets the time, you are going to get the clock back! To verify your adjustment, install the minute hand, washer and pin. Allow the clock to reach the hour under its own power to get a true reading on the strike point. If the movement will run with the pendulum off, it will speed up the process of waiting for the hand to reach the hour. Not all movements will run this way.

The Hamburg-American movement has four springs to assist the strike train. I have already mentioned the spring on the hammer arbor, which helps to produce a loud gong sound by reinforcing the force of gravity on the hammer. Another spring operates on the arbor carrying the count, drop and unlocking levers. This one is necessary for positive locking action on the strike train. A third spring is mounted above the warning lever to assure that the lever snaps smartly downward at the hour and half hour to begin the strike. You can see this spring in Figure 38 (top). The fourth spring is really the acting part of the lifting lever; it provides the "give" when the minute hand is turned counterclockwise through the strike point.

An eight-day test completes the repair of this clock. Listen occasionally for the correct strike count and make any necessary adjustments. Remember that the count wheel strike mechanism cannot correct errors which occur. It is important to let the clock owner know about this characteristic. To correct the count, a cord attached to the unlocking lever is pulled to start the strike. After the strike is completed, the mechanism can be "tripped" again, if necessary, until the strike agrees with the hour hand position.

9

FRENCH COUNT WHEEL

From a collector's standpoint, French clocks are hard to resist. Besides being elegant, they are still reasonably priced when compared to other high quality clocks. This chapter covers the common type of round French movement which strikes the hours and half hours on a bell mounted to the back of the movement. It appears in this section of the book because a count wheel controls the striking. A similar French movement with rack strike is featured in Chapter 12.

The plates of the round French movement are less than 3-1/4" in diameter. The pivots are often as small as .015" in diameter and are hardened in most cases. Care is required to avoid breaking or bending them. Repivoting can be difficult because of the small sizes and hard steel. Beginners should work on sturdier movements until skill and confidence are gained in handling small parts.

The French movements are often brought to the repairer because dirt and wear have combined to bring the mechanism to a halt. Mainspring condition is especially important. A fine French striking clock, recently taken apart for cleaning, illustrates this point. It had been repaired within the last year or so, but it stopped repeatedly. From the outside, the movement appeared relatively clean and usable. The back plate should have been polished, but that would not have been a factor in the performance of the clock. Complete disassembly revealed the problem; both mainsprings were dirty, covered with rancid old oil, and stretched into a "cone" shape from being pulled from the barrels with pliers. Both springs had to be replaced.

Fig. 39. French striking clock, black slate case 12" wide by 8" high; movement by Japy Freres et Cie.

Fortunately, the clock illustrated in Figure 39 had better mainsprings. The movement was dirty, although it would run. The main defect was the worn-down pivot at the second arbor rear location, in the time train. This is a common place to find wear in the French movements.

Disassembly and Inspection of Parts

Let down both mainsprings and proceed to disassemble the movement for cleaning. Look for bent pivots and damaged wheel teeth, which will have to be repaired. Unscrew and remove all cocks and bridges; cleaning solution will seep under them if they are left in place. For the same reason, take off the warning lever and the unlocking lever (Figure 42). Remove the mainsprings from the barrels and check them for cracked ends.

Inspection of strike parts should include several items, which can be located on Figure 42 and 43. Make sure the count lever is not loose. All pins on the pin wheel should be straight and tight. They can be carefully straightened if necessary. The locking and warning pins should also be checked. The wearing surfaces of the warning lever should be polished if they have become grooved. The hammer head must be tight on the hammer shaft. Locate any worn pivot holes that will require a bushing. Polish the pivots, or at least check each one and decide whether it is smooth.

Fig. 40. Front view, French count wheel movement

Reassembly

The reassembly process worries some repairers, because this is the phase of the repair which can result in broken pivots. It is also the time consuming part of most French clock repairs, for the difficulty lies in all the adjustments and small details.

In most cases it is a good idea to reinstall, at this time, the cocks for the center arbor (rear) and third time arbor (front). Leaving them off doesn't seem to make the reassembly process easier. Do leave off the cock for the pin wheel arbor (rear), which is shown in Figure 42. It is put in place later, as strike adjustments are made.

The center arbor and wheel must go into the front plate first. Then add the two barrels, which are not interchangeable. Some parts are marked, either at the factory or by repairers. Factory markings are more reliable, usually consisting of a pair of punch marks on certain strike parts, such as the barrel and cover, and a single punch mark on time train parts. (The situation is somewhat confused by the single index marks often placed on strike wheels. Although these are supposed to be lined up with marked pinion leaves in some movements to assure correct indexing of wheels, I do not use them myself. In any case, the marks are not completely consistent on all movements.)

Add the other arbors into the front plate, but leave out the pallets, and the back cock which supports them, until the plates have been pinned to-

Fig. 41. Rear view of movement with bell removed (left); rear quarter view showing strike parts, including the bell.

FIG. 42. FRENCH MOVEMENT, COUNT WHEEL TYPE

STRIKE PARTS ON FRONT AND REAR PLATES

WINDING MECHANISMS, HOUR WHEEL AND TIME TRAIN PARTS NOT SHOWN

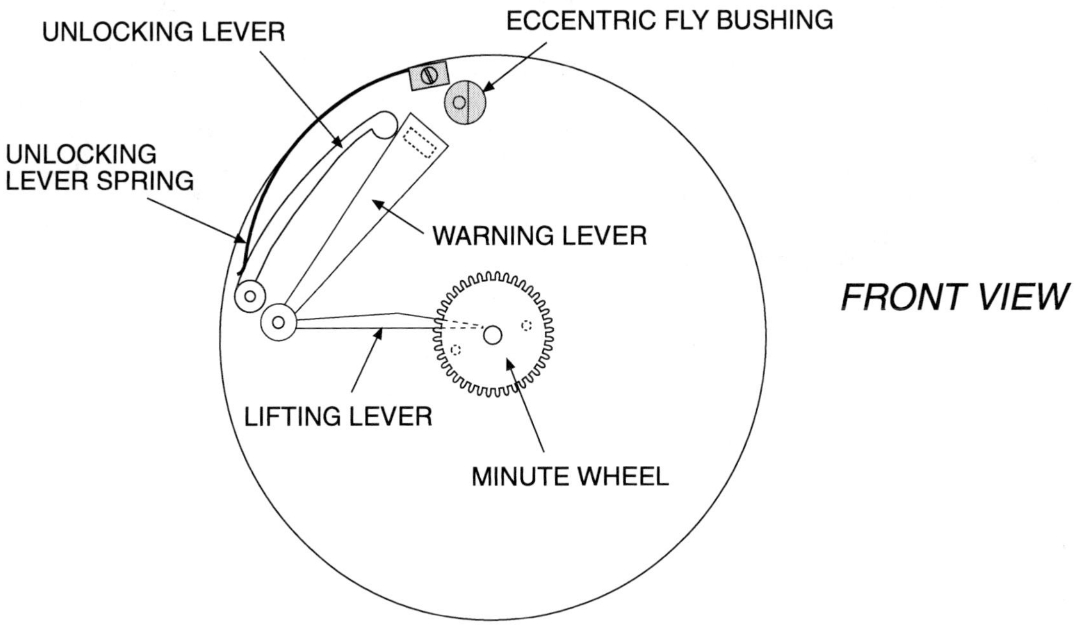

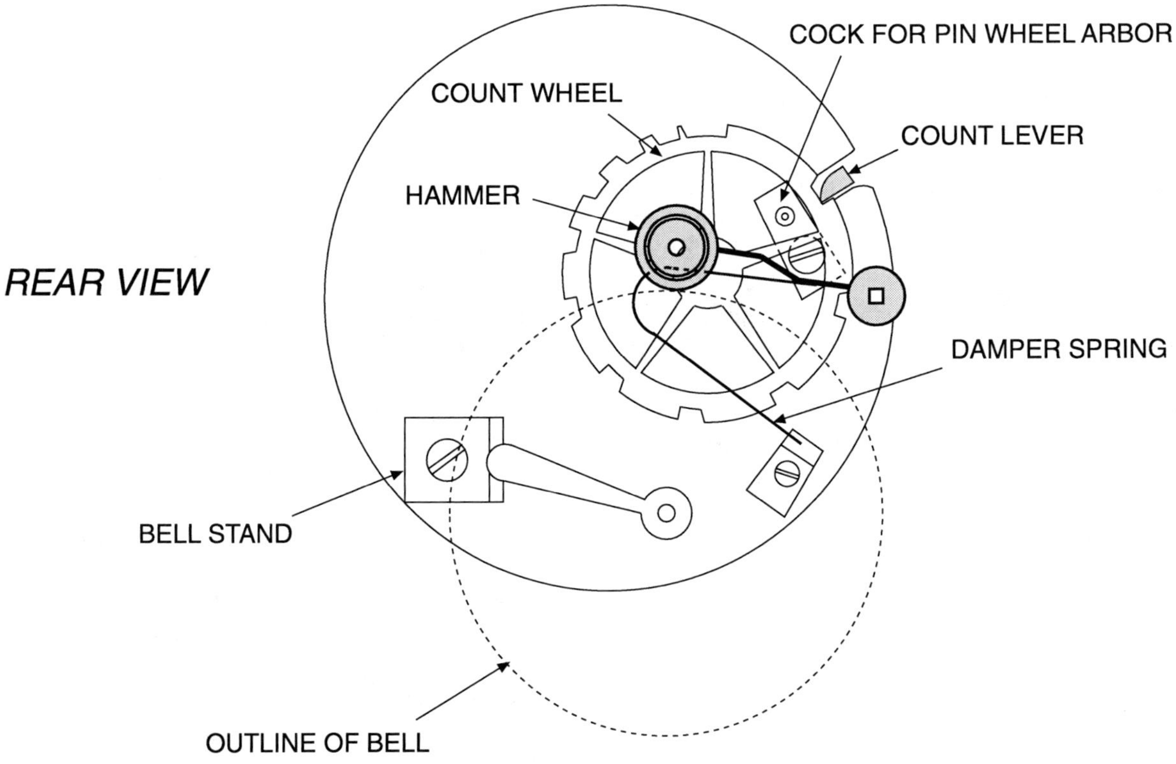

gether. Use care, and don't force pivots into place. Tweezers are suggested for moving pivots safely into the holes. Add the hammer arbor and the arbor which carries the locking lever. Install the taper pins to hold the plates together. These pins should be steel, not brass, and should be cut carefully to length and the ends neatly filed, before they are pressed into the pillars.

Finish this stage of assembly by adding the unlocking lever and the lifting/warning lever to the front of the movement. Install the count wheel on the back of the movement. Do not wind the mainsprings at this stage.

Warning Adjustment

Bring the wheels around so that the locking lever rests against the lock pin. Now check the position of the warning pin. In this movement, the pin should be oriented in a 6 o'clock position to create about a half turn run of the warning wheel and pin, when the warning occurs. To make this adjustment, remove all the taper pins from the pillars, except the bottom one. Carefully ease the plates apart enough so that the pinion on the warning arbor can be slipped out of engagement with the locking wheel. Move the warning wheel and then fit the plates together again. The fly pivots will be out of the holes, so be careful to fit them in again.

Pin Wheel Adjustment

Since the cock for the pin wheel rear pivot was left off until now, it will not be necessary to separate the plates to adjust the indexing of the pin wheel. Even without the cock, it is possible to put the front pivot in the hole and hold the rear pivot parallel with a fingertip. With the lock pin still in place against the locking lever, check the position of the pin wheel. The hammer tail should be resting on its stop, not halfway lifted by one of the ten pins on the pin wheel. To correct the position of the pin wheel, slip the arbor backward to disengage the wheel from the pinion on the locking arbor. Somewhat by trial and error, fit and refit the wheel and pinion until the hammer tail is at rest and not touching any of the pins on the pin wheel. If the wheel is indexed one tooth to either side of the right place, it will not be correct. It can take a few minutes to make this adjustment; it is easier to see what must be done than to actually do it! Screw the cock for the pin wheel arbor in place when the adjustment is correct.

Count Wheel Setup

At this point, the clock could be run through some strike sequences. It is important for the count lever to stop at one end of the count wheel slot at the end of an hour strike, then stop at the opposite end of the same slot at the end of the half hour strike. If the count lever does not line up, do not bend it. Remove the count wheel and try it in another position. It is mounted on a square, so there are four positions in all.

Finishing the Assembly

Add the hammer and bell to the back of the movement (some movements may be gong strike). Bend the hammer shaft to obtain the clearance which results in a clear tone from the bell or gong. Do not put any pressure on the bond between the hammer shaft and the hub which fits on the square-ended hammer arbor. It may come loose and can be difficult to solder neatly back in place.

The movement featured in this chapter has a damper spring, which can be adjusted to support the hammer at the bottom of its stroke. This reduces the bouncing of the hammer and improves the tone as a result.

Front movement parts will also include the cannon pinion, which will carry the minute hand; the hour wheel; the minute wheel; and the cock which provides the front pivot for the minute wheel. It is not necessary to index the hour wheel and the minute wheel pinion in a special way, as is required for the rack strike version.

Install the pallets and back cock, then check the tightness of all pins and screws. In addition to normal oiling, lightly grease the lifting lever where it is touched by the two lifting pins (shown as dotted circles in the Figure 42 front view). The hammer tail can also be greased where it is contacted by the pins on the pin wheel.

Testing

Run the movement on a test stand or in the case. A problem in the strike will show up as an incorrect count (since the count wheel cannot correct itself), as a failure to strike, or as continuous striking.

If there is a problem, double check for loose, bent or rough pins, whether they be on the pin wheel, locking wheel or warning wheel. Perhaps the count lever is loose or bent. A rough surface on the warning lever should be polished.

Thoughts on the French Movement

Although the beginner should not start right out on French clocks, there is no reason to avoid them for too long. One can always buy a scrap movement and gain experience at any time, but the hobbyist should probably avoid for a year or two the disassembling of French movements which are heirlooms or clocks belonging to customers. However, once the skill in handling small pivots has been developed,

FIG. 43. FRENCH MOVEMENT, COUNT WHEEL TYPE

EXPLODED VIEW OF STRIKE PARTS

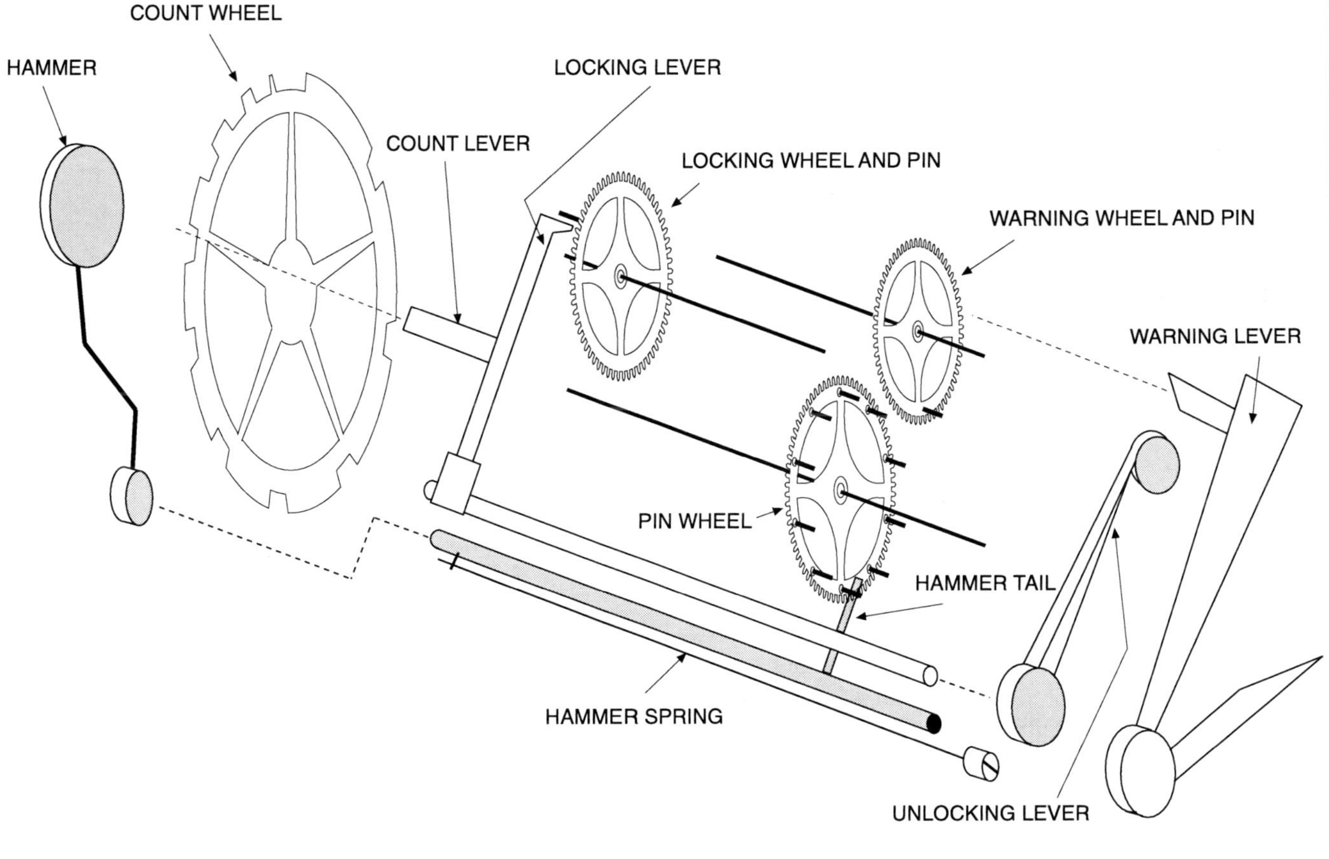

it is time to get to work on French movements. Except for the small size of the parts, a French striking movement is easier to assemble than an American movement, which has all those awkward, interlocking soft wire levers.

Fine points of finishing are important for French movements. When the movement is disassembled, thoroughly clean all wheels, pinions and other parts. Polish at least the rear plate with brass polish or diamantine powder and oil, then clean the plate thoroughly again to remove grit and polish. Each screw should be checked. If it is in good condition, it can just be cleaned. If the slot is damaged, then file it clean again, polish the screw, and "blue" it again. All taper pins should be replaced unless they are perfect; most will not be found in good condition. Always file the cut ends of taper pins, instead of leaving them rough.

Mainsprings are a leading area of neglect in French movements. It is possible to remove most of these relatively small springs and insert them again without a winder, but a winder is essential for a serious hobbyist or professional.

The clock hands make the first and most lasting impression of the condition of the clock. Hands should not be left rusty and dirty. Polish and "blue" them, either by using heat or a chemical made for that purpose.

10

SETTING UP THE RACK & SNAIL STRIKE

The rack and snail strike has at least one great advantage over the count wheel strike movements that were covered in Chapters 1 through 9. That advantage is its ability to maintain a correct strike count, at least within limits.

Consider the count wheel, which just counts one hour (or hour and half hour) after the other. If the clock is allowed to run down, or if the hands are turned rapidly, the count will be wrong when the clock is started up again. A count wheel mechanism that last counted 10 o'clock will resume by striking 10:30 or 11:00 next, depending on the design. That's why so many of the old count wheel clocks have a trip wire hanging down from the count lever. The clock owner can make the clock strike hour after hour, in order, until the strike agrees with the hands.

In comparison, the rack and snail mechanism can be allowed to run down, or it can be reset, usually without consequences. If the clock resumes striking, it will be able to count the correct hour as indicated by the hands. The key to this mechanism is a special cam, which is called the snail because of its shape (Figure 44). Each step on the cam represents a particular hour. Since the snail is attached to the hour wheel, and therefore geared in with the motion work, the snail is always ready to control the number of hammer strokes.

There is more variation in the details of rack and snail strike mechanisms than in the count wheel type, although there is still a single principle behind each of them. It is very helpful to have a set of instructions at hand for each individual rack and snail strike movement to be repaired, and the chapters which follow give details on several. Unfortunately, it's not possible to cover them all. The repairer must learn the concepts which apply rack and snail mechanisms in general. Whatever the variation, certain basics will apply. This chapter covers the main points to be watched in all rack strike movements.

Fig. 44. The snail

Hammer Tail

There will be a pin wheel of some kind in the mechanism. It can be a stamped-out "star" which lifts the hammer tail with each point, or it can be a set of pins driven into a wheel. The number of pins varies from two to ten or more. It's a rule (almost always true) that the hammer tail should be at rest and not left partly raised by a star wheel or pin, when the clock is not striking. The reason: a strike train that starts up with the hammer under load may stall.

Figure 45 shows a New Haven rack striking movement as an example. The hammer tail is clear of

FIG. 45. NEW HAVEN RACK AND SNAIL STRIKING MOVEMENT.

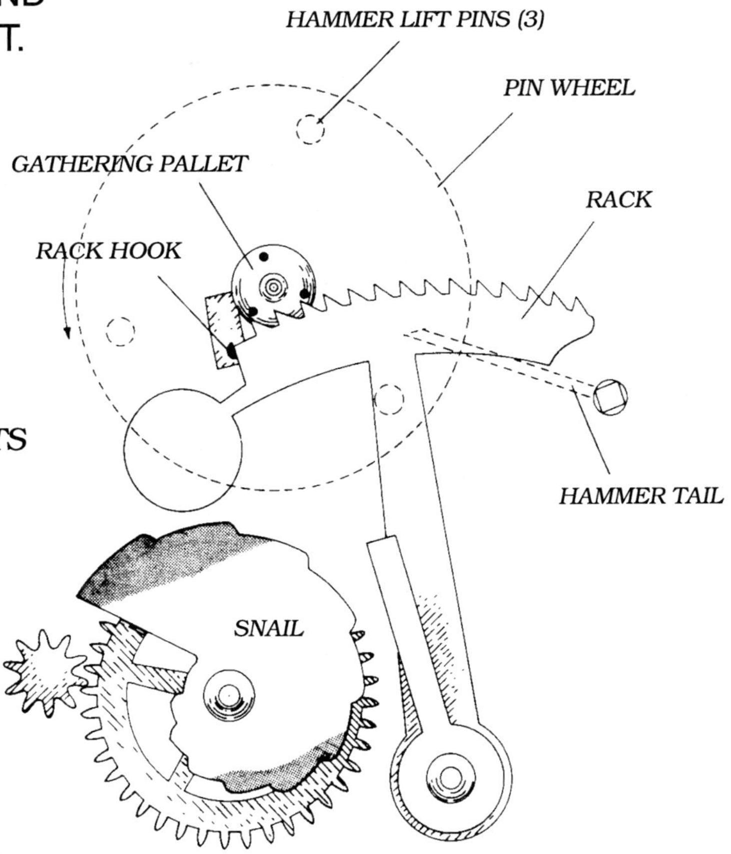

REAR MOVEMENT STRIKE PARTS

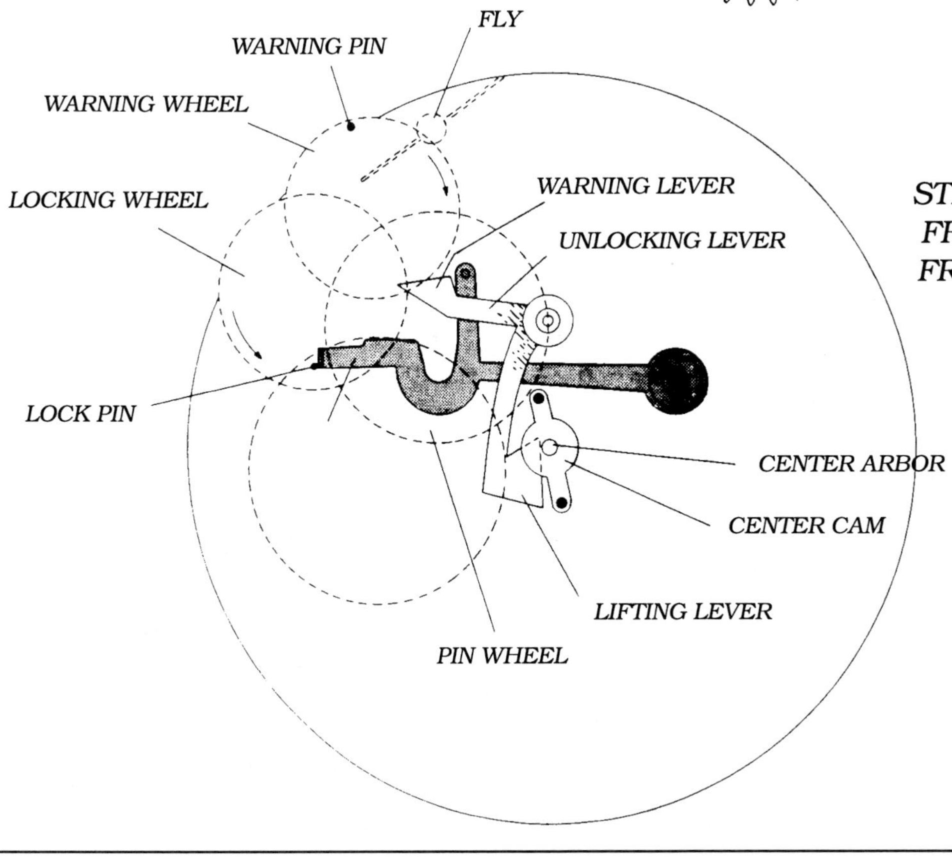

STRIKE LEVERS, VIEWED FROM THE FRONT WITH FRONT PLATE REMOVED

the three pins on the pin wheel. In some movements which have a star instead of individual pins, the star is held onto the arbor with a stiff clutch that allows movement. In most cases, however, the movement plates must be separated slightly to disengage the wheel. It is then indexed and tried again until the hammer will remain free of the pins or star. During this test, keep the lock pin against the locking lever.

Warning Action

There are two general ways for the warning to operate. Warning in a rack and snail movement has the same purpose as in a count wheel movement: the train is "set" or made ready to begin striking exactly on the hour or half hour. To accomplish this, the locking of the train is transferred from a slow moving wheel to a faster moving wheel, higher up in the train. Warning occurs about five minutes before the hour in many rack and snail movements, but happens as little as two minutes before the hour in some.

In the New Haven example in Figure 45, the warning pin is set so that it will run about a half revolution until it is stopped by the warning lever. In this way, the locking action is transferred from the lock pin to the warning pin; that is, it is passed from a slow moving wheel to a faster moving wheel. In such a movement, the warning pin position must be set by assembling the movement correctly. As in the hammer tail adjustment, the strike train is placed in the locked position, with the lock pin against the locking lever. It will be necessary to separate the plates enough to disengage the wheel (normally the pin wheel) from the pinion on the gathering arbor. Mesh the two so that the warning pin will be positioned for about a half turn of warning "run". Some movements may be ideally set for more or less warning run, but a half turn of the warning wheel could be considered as suitable for most movements.

Other movements use a single pin for locking and warning. The Seth Thomas No. 120, featured in Chapter 13, is such a movement. There is no need to "set" the warning, since the lock pin moves from the locking to the warning lever when the warning occurs.

It would be logical to ask why the movement should not simply be put together correctly in the first place, with the hammer tail and warning pin set up. The answer is that it is difficult to assemble so many arbors and levers between two clock plates, all the while making sure that pivots are not forced or bent, and still be able to control the precise indexing of wheels and pinions. Sometimes the pieces fall together the right way the first time, but in most cases at least a small adjustment will be required.

Gathering Pallet

The gathering pallet is a part which picks up, or gathers, rack teeth to perform the counting function and control the strike. The rack hook holds the rack teeth as they are counted off. It's almost like having 12 apples in a basket. At 12 o'clock, all 12 are dumped out at once. Then they are picked up, one at a time, and placed back in the basket. At one o'clock, only one apple is tipped out of the basket, then put back. Two o'clock involves two apples, and so on.

The shape of the gathering pallet varies from movement to movement. Two basic styles are shown in Figure 46. The round gathering pallet, on the left, is a brass disk with three pins; some pallets have only one or two pins. This style simply gathers

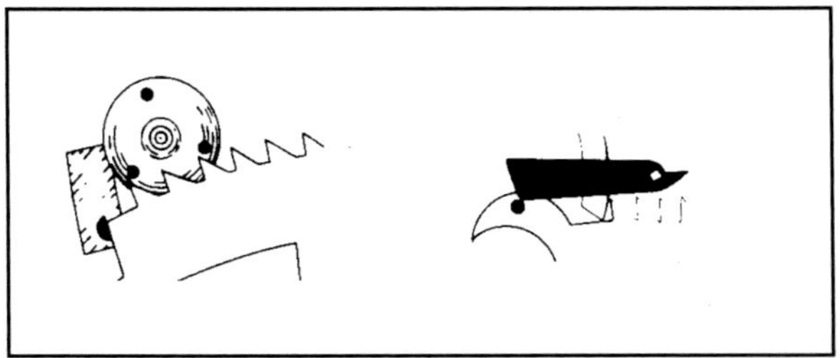

Fig. 46. Two styles of gathering pallet

the rack teeth, which are then held by the rack hook. The other style, shown on the right, gathers rack teeth with one side and locks the gear train on the other side. A third type is a combination of the two styles; it is a kidney shaped piece which functions as a cam to lift the rack hook at each strike, and it also has a pin for gathering teeth.

It would be difficult to cover all possible adjustments of each type of gathering pallet here. Specific adjustments are covered in each chapter in this section on the rack and snail strike. Generally, however, gathering pallets similar to Figure 46 (left) are adjusted so that the pin or pins (if that is the style) do not interfere with the dropping of the rack onto the snail at each hour. The rack cannot count correctly if it is not allowed to operate with the snail. Gathering pallets similar to Figure 46 (right) have a locking function, and often are fitted to a square arbor. This limits the number of possible positions on the arbor to four, and tends to simplify gathering pallet questions.

Sometimes, repairers become frustrated in trying to adjust the gathering pallet position on its arbor. In some clocks, the gathering pallet position even sets the rest position of the hammer; in others, adjustment is only a matter of keeping the gathering pallet pins clear of the rack teeth when the train is at rest. The final adjustment may be a small one. Be careful to make sure the gathering pallet is loose on the arbor before you twist it. Otherwise, the arbor may snap off!

One hint on movements with the round gathering pallet: if there are problems, check the locking function of the strike train. If the lock pin is too close to the locking lever when the rack hook drops off the end of the snail, signalling the end of the strike sequence, the locking may be inconsistent. The pin might catch the lever one time and miss it the next. The extra turn of the gears will cause a gathering pallet pin to remain engaged with the rack. The clock will not strike the correct hour the next time. Re-index the locking wheel for a positive action. More information on this adjustment is contained in Chapter 13.

Rack and Snail

A simple adjustment to make the rack and snail work correctly together is often overlooked or done carelessly. Figure 47 illustrates a snail and hour wheel which have been incorrectly set within the motion work of the clock. The arrows are placed 30 degrees apart to indicate the pattern of points the rack tail will touch each hour. The 30 degree spacing divides the circle of the snail into 12 equal parts, one for each hour. On this particular design, the arc of the "1" step is just long enough to permit the rack tail to fall upon it two hours in a row. That leaves us with a single strike at 12 o'clock and all the other hours correct, although the rack tail just barely catches the edge of each segment on the snail. In some other clocks, the one o'clock segment of

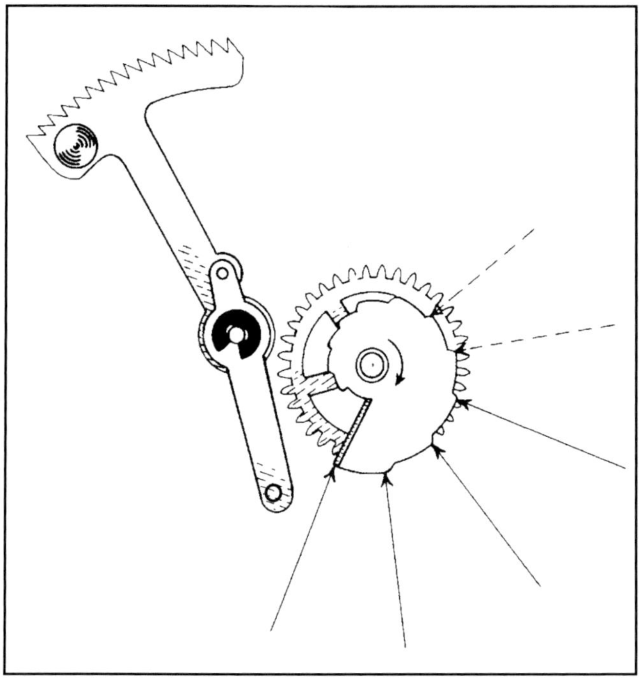

Fig. 47. A rack and snail, improperly set up as shown, will strike one or more hours incorrectly or jam the clock.

the snail is shorter. Instead of landing on the 1 o'clock plateau at 12, the rack tail jams halfway down the side of the snail, between 12 and 1, stopping the clock.

Figure 48 shows the parts involved with this adjustment. A split washer or pin holds a washer over the minute wheel pinion. Sometimes the fastener serves as fastener and washer. The washer keeps the hour wheel engaged with the pinion.

To make the adjustment, remove the washer and pull off the hour wheel so that it can be indexed differently with the pinion. In Figure 48, the snail is

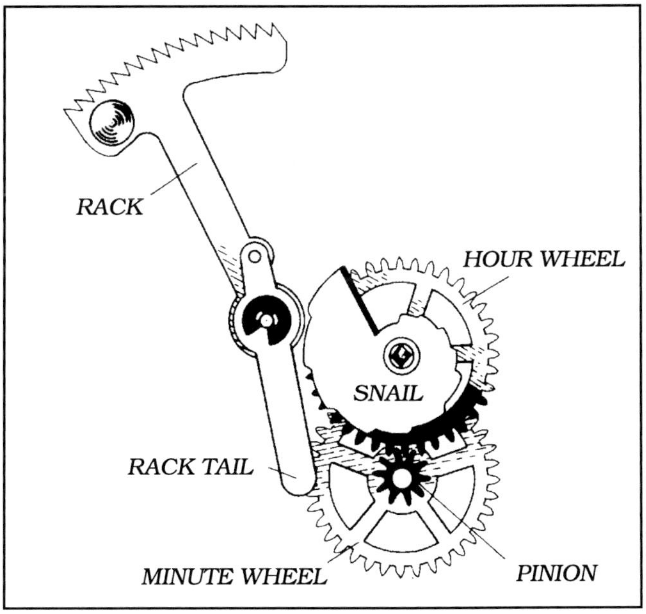

Fig. 48. Parts involved in adjusting the rack and snail

oriented, for clarity, so it does not cover up the rack tail. For adjustment purposes, install the snail in a trial location at 12 o'clock; the rack tail should fall to the lowest point on the snail. The 1 o'clock position will also be checked as described below.

Figure 49 shows the two critical settings for 12 and 1 o'clock. If both of these are correct, all the other hours should strike correctly. Only damage to the snail itself would cause a problem with one of the other hours. Such damage is not common in modern clocks or even in the mass produced clocks made since the middle of the 19th century. It is more often seen in antique clocks over a hundred years old. Old, handmade clocks are more likely to have design errors, and may have suffered damage through the years. Our purpose here is to adjust a snail that is not defective.

In Figure 49 (left) the rack tail is at the 12 o'clock location on the snail. Turning the minute hand to the hour should produce this result. If the adjustment is incorrect, the rack tail will jam between the 12 and 1 segments of the snail. Pull off the hour

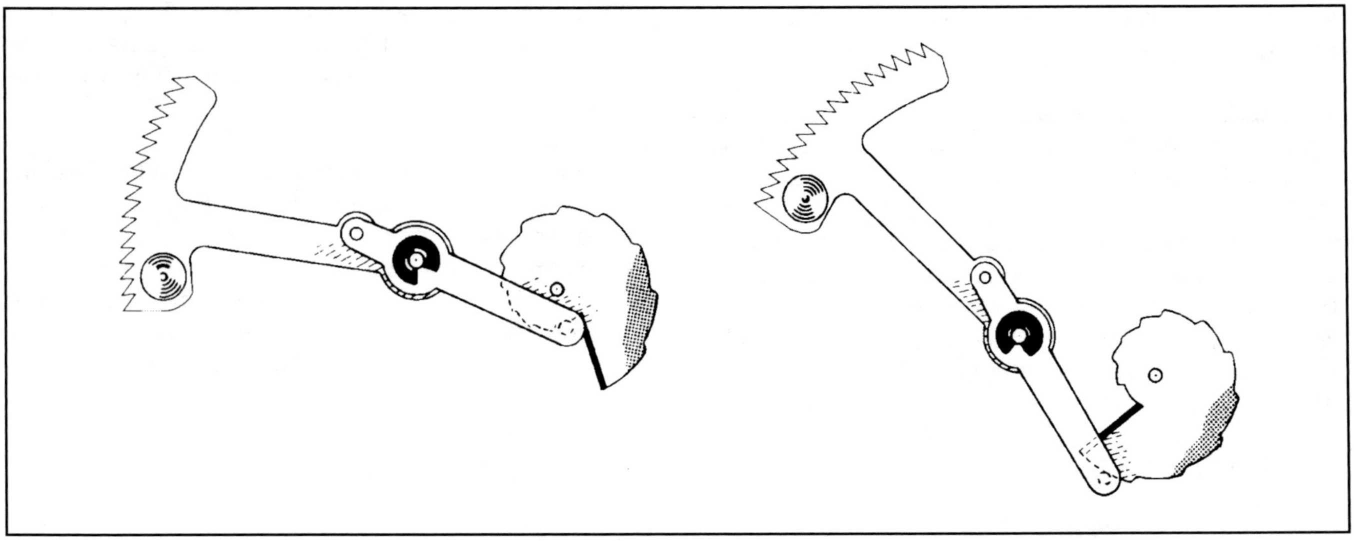

Fig. 49. The 12 o'clock setting (left), showing the rack tail on the snail. The 1 o'clock setting is shown on the right.

wheel and snail, then re-index the parts so the rack tail is placed correctly for 12 o'clock.

It is just as important to check that the rack tail will fall onto the 1 o'clock segment at the next hour. Figure 49 (right) shows this position. If the rack tail must fall near the edge of the step, it doesn't represent a problem, as long as it lands safely.

Some snails are made in such a way that the best "12" adjustment won't allow the rack tail to reach the "1" step; it falls just short and jams there. In that case, you would move the hour wheel and snail one tooth in the counterclockwise direction, then mesh the wheel and pinion again. Although the rack tail would crowd the edge of the "12" step, it would still work at 12 and 1 o'clock. With most clocks, the best adjustment allows the rack tail to touch the snail in the middle of each step.

In many rack and snail setups, there is only one correct way to mesh the hour wheel and the pinion for correct striking. Any deviation will cause an incorrect strike on one or more hours. Other clocks are built with enough room for the adjustment to fall one tooth either way without a malfunction.

11

ENGLISH BELL STRIKE

Anyone who works on clocks will be asked, before very long, to repair one of the old English bell strike grandfather clocks with rack and snail strike on the hours. The job is certain to be a test of skill. Although the movements in these clocks are of basically similar pattern, they vary greatly in the nature of the repairs required.

The movements are simple and well constructed. A century ago, it was probably easy to repair most of them! Unfortunately, the years have taken their toll. Two factors make present-day repairs more difficult: wear from many years of timekeeping; and defects carried over from many repairs, some of which were badly done. A poor repair becomes *your* problem, even if it was made generations ago.

A sampling of these movements will show all kinds of damage. Broken wheel teeth, cracked wheel castings, worn pinions, loose and rusted levers, worn pivots and holes, cracked gathering pallets, grooved pallet faces, bad clicks, and heavy-handed solder repairs are some of the main areas of damage.

The most difficult part of the repair is not the cleaning of the components or the reassembly of the movement. Achieving long-term reliability is the greatest challenge. A dozen parts threaten to work loose, split, hang up, or otherwise fail. The repairer should look carefully at the entire movement before quoting a price for repair.

The drawings in this chapter show how simple the English bell strike mechanism really is compared to most other clocks we repair. It's so easy to assemble one of these old movements, but fitting the pieces together is just the beginning. Accumulated wear combines with the scars from old repairs. What you are left with is a movement that can fail in many ways. More than once I have seen them work well on the test stand, only to act up soon after delivery to the owner.

Strike Train Operation

Figure 50 (next page) shows the strike parts on the front of the movement. This rack and snail mechanism has a heavy gathering pallet (7) which locks on the rack pin (6). The cannon pinion (17) rotates clockwise with the center arbor (1). It is held by a clutch arrangement allowing the owner to turn the hand (and the cannon pinion) to set the time, or to make the clock strike. The cannon pinion turns the minute wheel (12) counterclockwise once per hour. The strike lift pin (11) starts the hour strike.

As the pin raises the strike lifting lever (10), the warning lever (9) pushes up on the rack hook (8), releasing the rack (5). The rack tail (15) falls to a certain spot on the snail (2) to determine the number of hammer blows to be struck.

At the same time, the gathering pallet (7) is free to move. The warning, which will be explained in

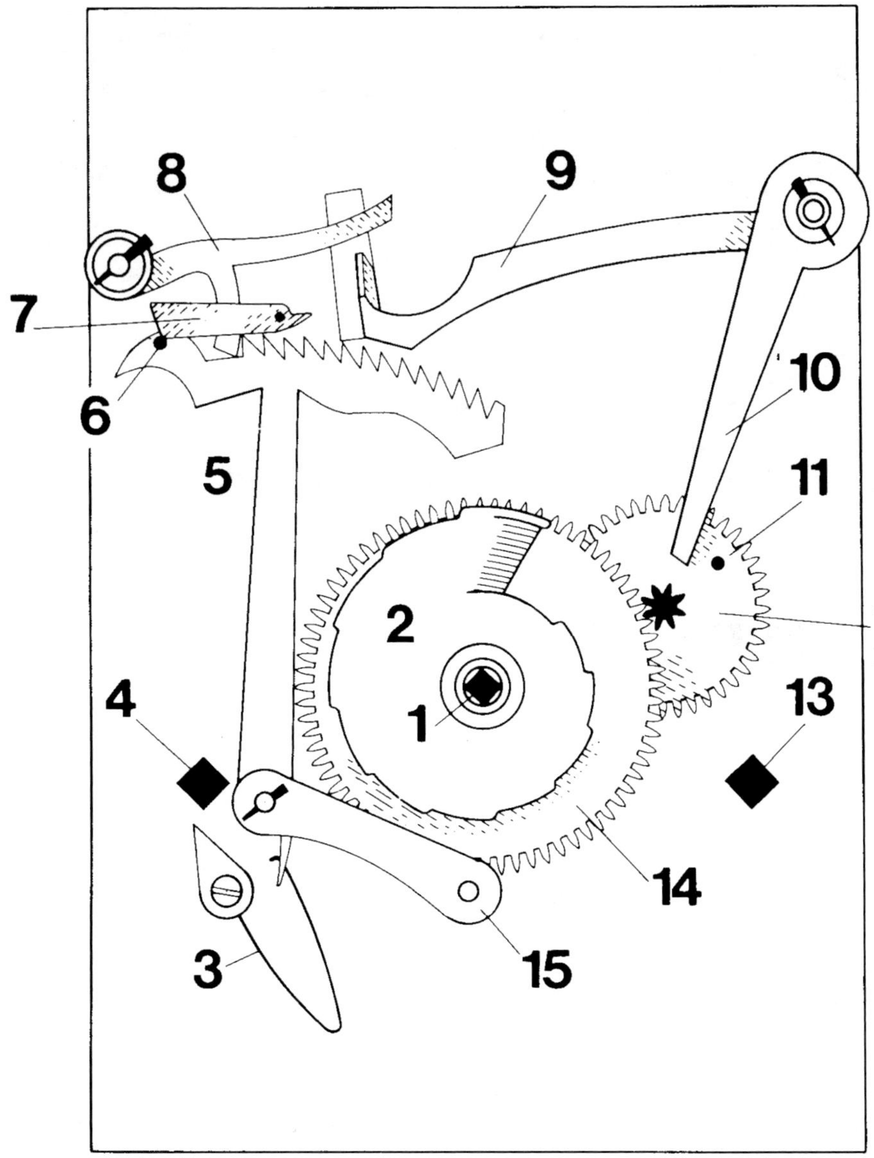

FIG. 50. ENGLISH BELL STRIKE
FRONT STRIKE PARTS

1 CENTER ARBOR
2 SNAIL
3 RACK SPRING
4 STRIKE WINDING ARBOR
5 RACK
6 RACK PIN
7 GATHERING PALLET
8 RACK HOOK
9 WARNING LEVER
10 STRIKE LIFTING LEVER
11 STRIKE LIFT PIN
12 MINUTE WHEEL
13 TIME WINDING ARBOR
14 HOUR WHEEL
15 RACK TAIL
16 HOUR BRIDGE
17 CANNON PINION

DETAIL OF LIFTING MECHANISM, SHOWN SMALLER SIZE

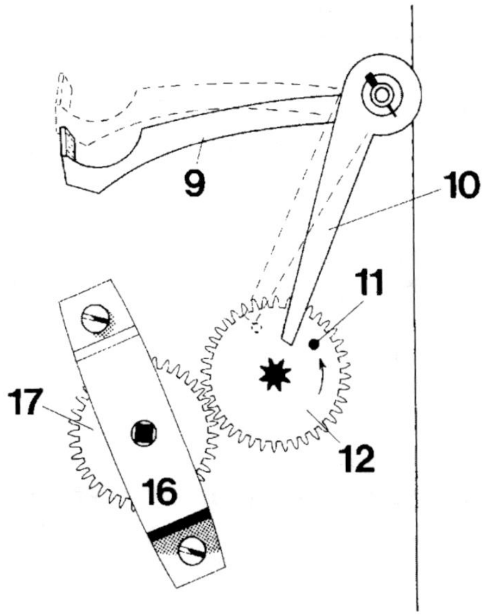

more detail later in this chapter, occurs now. The warning lever stops the warning pin. At the hour, the strike lift pin rotates far enough to drop the strike lifting lever. The corresponding drop in the warning lever releases the warning pin. The strike train starts up and continues to run, as the gathering pallet counts off rack teeth. Finally, the cycle is completed as the gathering pallet locks on the rack pin.

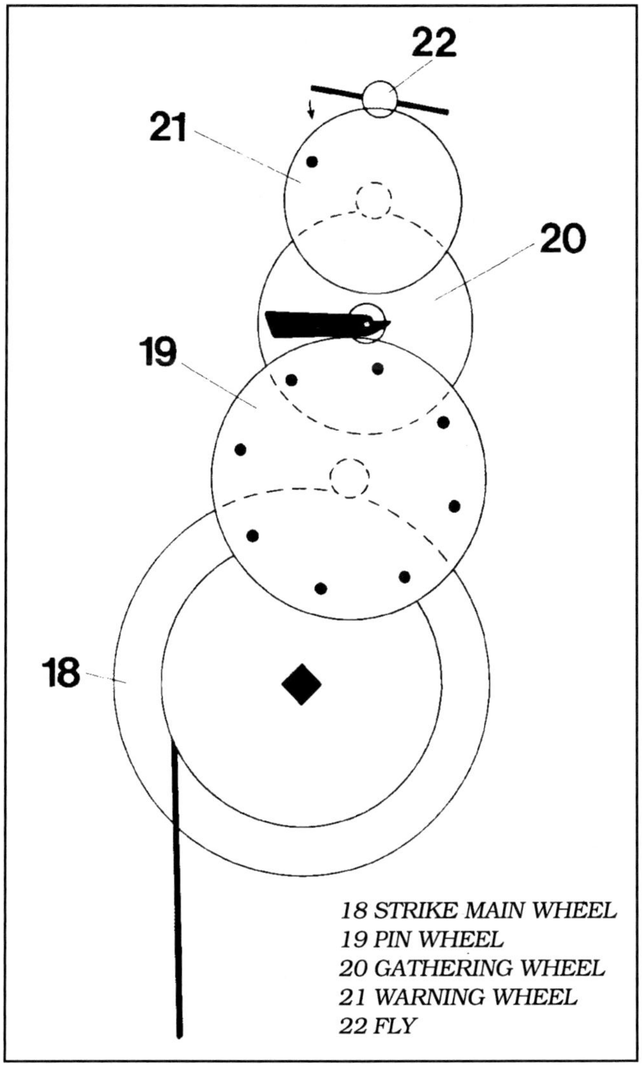

Fig. 51. Strike train layout

Strike Train Layout

It is worthwhile to look at the general layout or arrangement of the strike train, which shows the simplicity of the design. Figure 51 illustrates the five arbors. At the bottom, the main wheel (18) provides power by means of a weight, usually about 13 pounds of cast iron. Next, the pin wheel (19) is fitted with a number of pins (eight in this example) which will lift the hammer tail. Moving up, we have the gathering wheel (20) and the warning wheel (21). Finally, there is the fly (22) for speed control.

Hammer Arrangement

Figure 52 shows the hammer arrangement. Imagine the front plate removed and that you are looking inside. The pin wheel (19) provides lifting action for the hammer (27), which strikes a bell (28). Hammer tension is supplied by the hammer spring (23). As the lifting pins raise the hammer tail (24), the hammer arbor (25) pivots, carrying the hammer with it. The upper left movement pillar (26) acts as a damper to help limit the bouncing of the hammer.

You can bend the hammer shaft to bring the hammer head closer to the bell or further away. If the hammer is too far away, the sound is faint, and if too close, it stays against the bell. Adjusted correctly, the metal hammer raps on the bell to produce a loud, piercing tone. It's a bit too much noise for many people, and you will see everything from tape to cotton balls wrapped around the hammer to deaden the sound. Make sure you satisfy the customer on the volume. Otherwise he may call you back to quiet the bell.

By adjusting the hammer spring (23), you can exercise some control on the noise level. Even when the movement is assembled, you can remove the spring and change the tension easily. But if the tension is too light, the strike sound will be faint and erratic. In addition, the hammer will bounce too much. As shown in Figure 52, part of the hammer rests on the flat top of the spring. A weak spring will not control the bouncing, so the hammer head will hit more than once. A stronger spring will limit bouncing, at the same time it raps the hammer firmly on the bell. But excessive tension will overload the strike train, inevitably causing it to stall. Of course, the bell itself has everything to do with the sound. Often the bell is not original or has been cracked and repaired. You may need to replace a bell to achieve a good sound. If the bell stand (29) has been bent over to accommodate a different diameter bell, you should be especially wary. It may be hard to adjust the hammer to hit the bell properly.

Strike Train Assembly and Adjustment

The drawings show what you should look for as you assemble the strike train. There are only a few basic adjustments to make. Figure 52 shows the hammer tail (24) at rest. It is free of the pins on the pin wheel (19). This means that the hammer does not end up on the rise at the end of any strike sequence. As the strike begins on the next hour, the strike train is able to reach full speed before a pin engages the hammer tail. The hammer load is considerable, so it is not a point to be taken lightly.

To observe this hammer action, assemble the movement including the gathering pallet, rack, and

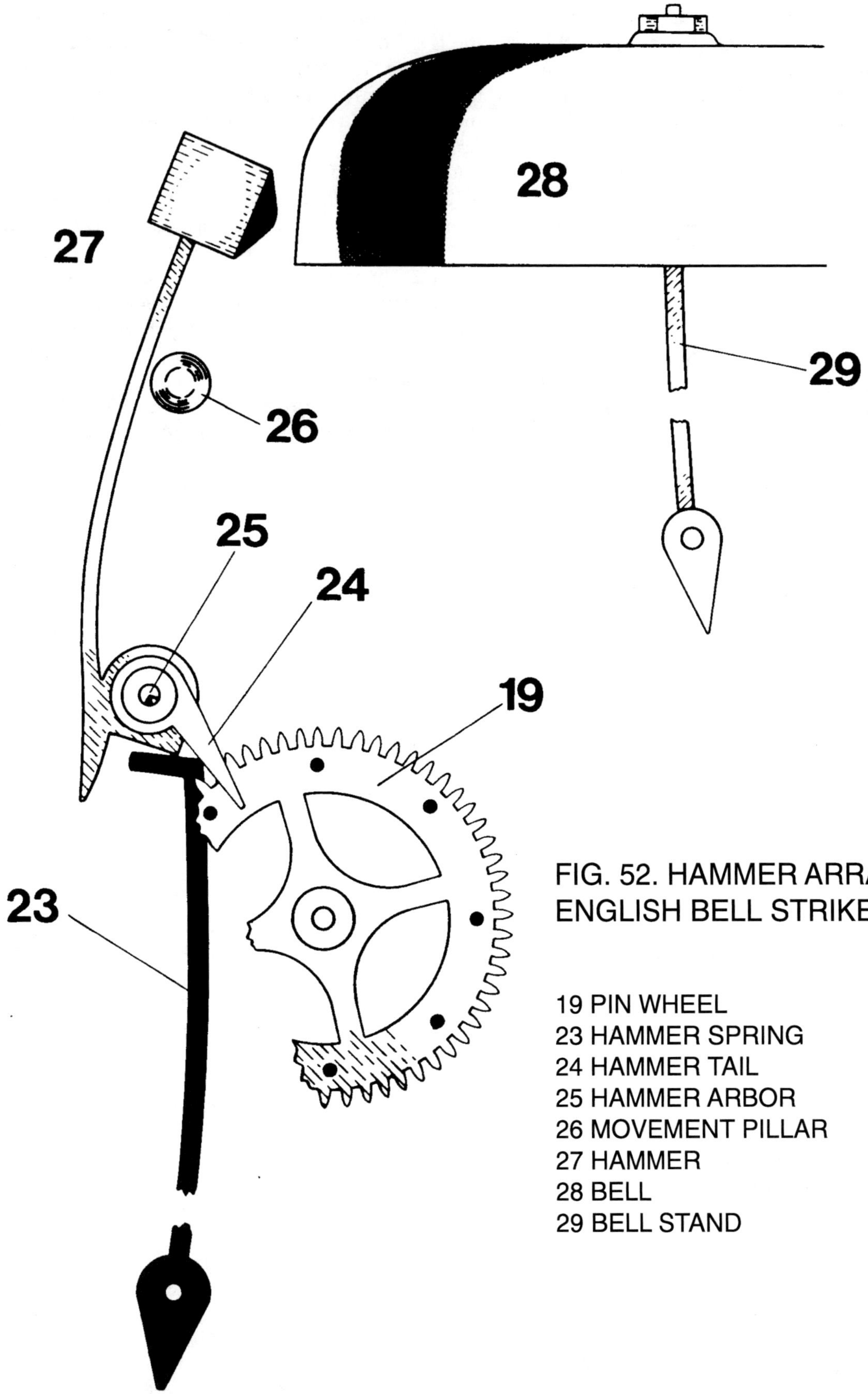

FIG. 52. HAMMER ARRANGEMENT, ENGLISH BELL STRIKE

19 PIN WHEEL
23 HAMMER SPRING
24 HAMMER TAIL
25 HAMMER ARBOR
26 MOVEMENT PILLAR
27 HAMMER
28 BELL
29 BELL STAND

rack hook. Rotate the wheels with your finger until the gathering pallet locks on the rack pin. Remember that you can install the gathering pallet four different ways on the square arbor. Each position will change the rest position of the hammer. The other way to make the adjustment (so the hammer is at rest when the strike train is locked) is to separate the movement plates slightly. Disengage the pin wheel from the pinion on the gathering arbor, then mesh the teeth in a different spot. Be careful not to bend the pivots.

The only other internal adjustment to check is the position of the warning pin. When the strike train is locked, with the gathering pallet resting on the rack pin, the warning pin should be seen at the 11 or 12 o'clock orientation, as viewed from the front. Figure 53 shows the warning wheel as a dotted circle, with the pin in black. The pin must be in this position to allow the warning action to work. To change the adjustment, disengage the warning arbor pinion from the gathering wheel below it. Mesh the gears at the correct location, then fit the plates together and check the warning action.

Pay particular attention to the lifting mechanism, which is shown in the detail view on the bottom of Figure 50. As mentioned earlier, the minute wheel carries the lifting pin, which starts up the strike each hour. By installing the minute hand on the movement, with rack and snail removed, you can turn the hand to determine when the strike will occur. The strike lifting lever should drop off the strike lift pin exactly at the hour. If the minute hand shows that the "drop-off" occurs early or late, an adjustment should be made. Loosen the hour bridge (16). You will see that the cannon pinion (17) is held under tension by a spring underneath it. This provides the stiffness needed to keep the hand in place. Pull up on the cannon pinion and disengage it from the minute wheel (12). You can now change the drop-off point by meshing the cannon pinion and minute wheel at the right place. Once the adjustment is done correctly, tighten the hour bridge again; the minute hand will now properly indicate the hour strike.

The rack and snail must also be adjusted. The rack tail must fall to the 12 o'clock position, not halfway between the 12 and 1 positions. Check the 1 o'clock position as well. To make an adjustment, refer to Figure 50. Remove the pin from the minute wheel post, then take off the rack, and pull up the hour wheel and snail. This disengages the hour wheel from the minute wheel pinion, which is shown in black in Figure 50. Mesh the gears one or two teeth one way or the other, then try the adjustment until the rack and snail work correctly at 12 and 1 o'clock positions. Finally, replace the taper pin (and a washer, if one is part of the mechanism) in the minute wheel post.

While checking the rack and snail in this way, it is a good idea to check all twelve hours for the correct count. Wear or previous filing on the snail may produce an incorrect count on one or more hours. Before you consider that the snail must be repaired, double check to see if changing the mesh of the hour wheel (and snail) with the minute wheel pinion will correct the hour which has the wrong count, without affecting the other hours.

The rack spring deserves careful attention. The strike train is designed in such a way that the rack does not fall onto the snail by gravity. Unless the rack spring pushes it, the rack will not fall. Reshape the spring so that it will move the rack in a positive, although not unreasonably forceful, manner. If the rack spring is too badly damaged, replace it. Rack springs are available from clock material suppliers.

Lock-Warning-Running

Lock is one of three phases, or positions, that should be checked whenever you adjust an English bell strike movement. Figure 53 shows the gathering pallet locked on the rack pin. In many clocks, the point of contact between these two parts is not positive. A gathering pallet that is worn, cracked, or the wrong size may glance off the pin. The pin itself may be worn or loose. Instead of locking, the pallet misses and causes the rack hook to pop up again. Striking will begin over if the rack tail falls back onto the snail. After the count is completed again, there is another opportunity for the mechanism to lock, but another failure is likely.

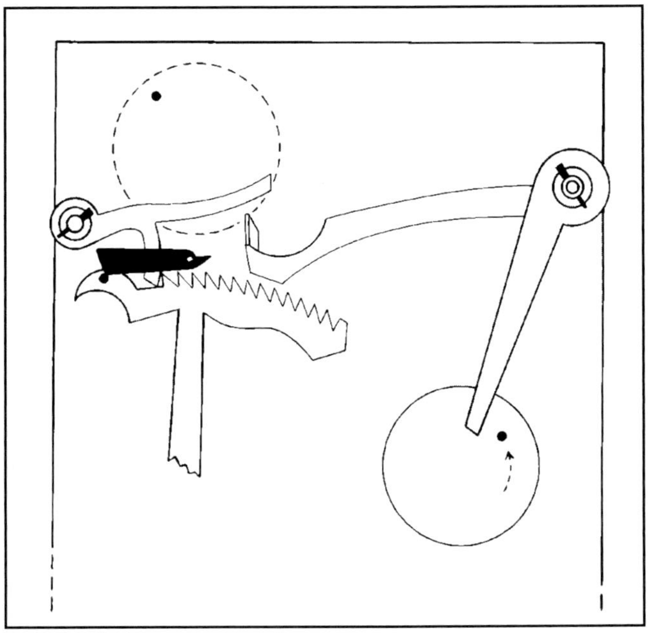

Fig. 53. Strike train in the lock position

The locking action is critical. The gathering pallet contacts the rack pin to stop the gear train. This produces considerable stress. Check the gathering pallet for looseness or cracks. If you need to make a new one, try starting with a blank purchased from a parts supplier. Whether you start from scratch or do it this way, it's a lot of work. Check the gathering pallet for smooth, dependable operation.

In addition, check the rack hook to see if it is part of any strike locking problem. A rough, worn surface will not slide smoothly down the last rack tooth. If the rack is sticking because of this, the rack pin will be to the left of the correct position, as the gathering pallet makes the last pass. The mis-locking often causes the strike to run on unpredictably. Look for rust on the rack hook post or an ill-fitting taper pin retaining the rack hook. Sometimes the post is loose in its threads and can easily be tightened.

Warning is the second phase, as shown in Figure 54. The warning pin, represented by a dotted circle which moves as shown by the arrow, has moved to the warning lever. The lever has, in turn, raised the rack hook, allowing the rack to move. Notice that the gathering pallet is free from the rack pin.

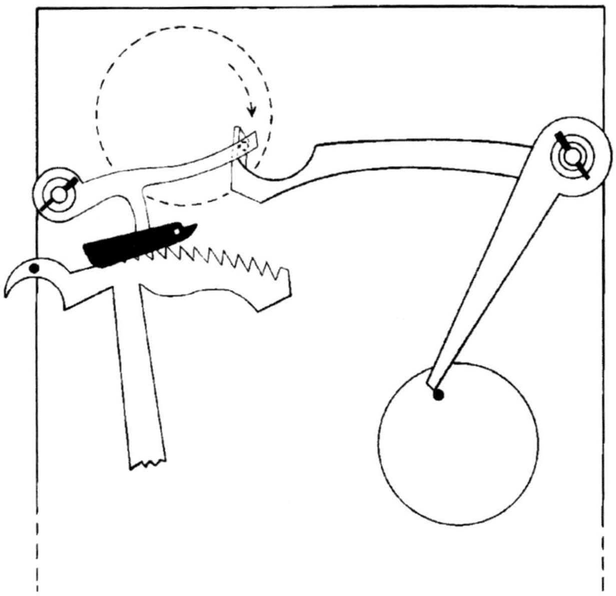

Fig. 54. Strike train in the warning position.

Check for rust or roughness on the warning lever post, and make sure the post is tight. Use new taper pins when installing levers.

Running (Figure 55) is the third phase in the operation of the strike train. At the hour, the strike lift pin drops the strike lifting lever (see the arrow), and the warning pin is free to rotate along with the warning wheel. If the warning pin does not slide easily off the warning lever, polish the lever and check for a worn or damaged pin. The clock will never work correctly unless you can count on positive action in the strike mechanism. A failure could happen every hour or perhaps just once a day, but the strike will fail nonetheless.

At the same instant the warning lever falls, the rack hook drops onto the rack. See whether the rack hook has dropped neatly into the space between two teeth. If the hook stops partly up the side of a rack tooth, the clock may strike the wrong count. This is often caused by the joint between the rack and the rack tail loosening over time, allowing the angle between them to change. Readjust the angle to allow the rack hook to seat itself properly between rack teeth, then tighten the joint. (Check to be certain the rack hook drops correctly at all 12 hours.)

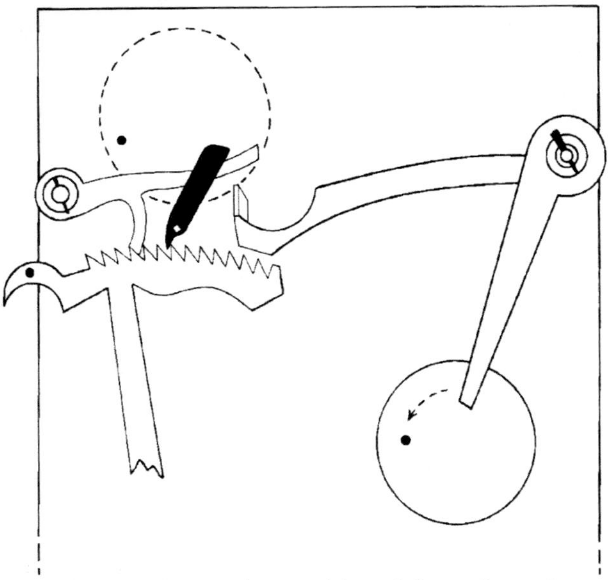

Fig. 55. The running position of the strike train.

As the strike train continues to run, the gathering pallet rotates, engages a rack tooth, and moves the rack one tooth space to the right, where the rack hook holds it in the new position. This action corresponds to one stroke on the bell. Watch the movement strike at 12 o'clock so you can spot any short teeth on the rack. The gathering pallet will release these short teeth too early, and the rack hook will not always seat itself properly at the next tooth. The result is an intermittent error in the strike count. If necessary, one or more replacement teeth can be fitted to the rack. In an extreme case, a new rack must be made.

More About a Loose Rack Tail

I mentioned above that the rack tail may be found to be loose on a movement. A brass bushing joins the rack tail to the rack. The parts must form a unit that is too tight to move with your hands. If you were to change the angle between them, the clock would strike too few or too many times each hour. A loose rack tail is common just because of the age of the English bell strike movements, but there is a more direct cause. The rack spring (#3, Figure 50) thrusts the rack tail against the snail every hour. The force is sufficient to break the bond between rack and rack tail after many years. Try tapping (riveting) lightly on the bushing to tighten the joint. Solder can be used, but only if it is applied properly and sparingly, so that it can hardly be noticed. The angle of the rack tail may have to be adjusted several times until the counting of all the hours is correct.

Testing

The English bell strike movement needs a relatively long test. A week of testing seems too short for a 150-year-old movement, when you consider that each part has some wear. Two or three weeks on the test stand will probably turn up any defects. If there are adjustment problems during the test, you need to keep the movement until it has been adjusted to work properly. If the customer becomes impatient, describe briefly your concerns and the need for complete testing. If he insists on "no more delays", it's for his convenience. Certainly it won't be for yours, if you have to return to the house.

12

FRENCH RACK STRIKE

This chapter is based on a French clock with a round, rack strike movement by Japy Freres. There are some differences in design from one French rack movement to another, but the clocks are similar and a repair procedure can be applied to them as a class. Figure 56 shows the china case for our subject, which has a scene painted on the front, beneath the dial. The case is made with a 3-3/4 inch opening in the front and rear. The round movement is pinned to the dial, and the two go in as a unit, through the front. Brass straps fasten the unit to the rear bezel. When you install the movement in the case, adjust the position of the dial so it looks straight from the front. Although many French clocks have an easily adjusted friction clutch on the pallet arbor, many do not. For the latter type, a fine beat adjustment can be achieved by loosening the brass straps and turning the dial-and-movement unit very slightly. If the clock is badly out of beat, this method will not be of any use because the dial would end up visibly tilted to one side.

Beginners are often put off by the small size (about 3-1/4 inches in diameter) of the round French movements. This is just as well because these movements are delicate compared to most other clocks. Small size means small pivots, and you must take care to avoid bending or breaking pivots during dismantling and reassembly. Just be extra careful with French movements, and don't pick one for your very first disassembly!

From the repair standpoint, French movements seem to present the same problems time and again.

Fig. 56. Rack strike clock by Japy Freres. The china case is 10-1/4 inches tall

They are subject to gummy deposits in the pivot holes, and these deposits will often stop the movements. The deposits may be abrasive mixtures of sour old oil, dirt, rust and moisture. The rear pivot of the time train second arbor is especially vulnerable in the typical French design. The pinion is at the end of the arbor, taking the full force of the barrel. The pivot may be badly scored, and a bushing is often required for this hole. In the strike train, the rear hole for the pin wheel arbor often shows enough wear to warrant a bushing. In general, a French movement does not need many bushings. It is more likely to need just one or two bushings badly, due to an isolated problem caused by dirt.

Disassemble and Clean

Disassembly begins with the letting down of the mainsprings. After this is done, remove the back cock and the pallets and take the hammer from the rear of the movement. Look at the leather hammer head and resolve to make a new one if necessary. Refer to Figure 57 to identify the front movement parts which are to be removed. Take off the minute wheel cock (15) and the minute wheel (14) along with the hour wheel and snail (11 & 12). Pry up and pull off the cannon pinion (13) with your fingers. The rest of the front movement parts can come off in any order, except the rack hook (5) must be taken off before the lifting and warning levers (6 & 7). To remove the gathering pallet (8), pry up evenly from two sides. Be careful with the gathering pallet, and don't try twisting it off. The danger is that you could break off the front pivot of the locking wheel arbor which carries the gathering pallet.

After you remove the pillar pins and rear plate, you might want to mark the barrels, so you can identify them easily later on. It is common, however, for the strike barrel, barrel cover, and arbor to be marked with a pair of dots (compared to a single dot for the time train barrel). If the parts can be identified as they are, then do not mark them yourself. Keep track of all the pins and screws, and the washers for the rack hook, rack, and hammer. Replace all bent or ill-fitting pins with new steel taper pins. Do any required bushing work and pivot polishing now. Check endshake and freedom of motion for all wheels. Insert the mainsprings back into the barrels and put on the barrel covers.

If you are going to clean a French clock, by all means take the time to remove and clean the mainsprings. Using a mainspring winder is certainly a good idea, so you won't stretch the spring out of shape by forming it into a helix. Replace any mainspring you find in this condition; it is not worth compromising your work to try to get by with marginal springs. If the end of the mainspring is cracked, it can be repaired in one of two ways. The mainspring can be shortened by about an inch, and a new hole made after the steel is softened. The alternative is to replace the mainspring.

It is not possible to describe each step of the cleaning process here, but a few tips will be helpful. Before cleaning the movement, you should remove three cocks, also called false pieces, from the movement. These are brass bars which support two of the time and one of the strike pivots. Cleaning solution becomes trapped under the cocks unless they are removed. Each cock has steady pins which will fit only in the correct location, and the holding screws are usually marked with identifying dots. A thorough cleaning should be followed by the polishing of at least the back plate of the movement. If the plate has taper pins driven into it, remove them. Polish in one direction with diamantine or brass polish; clean thoroughly again, including the pivot holes. The plate may be lacquered or left clean but should be handled from then on with gloves.

Begin the Reassembly and Adjustment

The next step is to get comfortable at a clean, well-lighted bench. Screw the two time train cocks back onto the plates but leave off the cock for the pin wheel arbor. Load all the arbors into the front plate, starting with the center arbor and the barrels, followed by all the other wheels. The hammer arbor and the adjacent arbor which carries the locking lever must be added in now. Don't install the pallets or the front movement parts at this stage. The main concern is to fit the plates together. Remember, the pivots are small and will break if they are forced. If a pivot will not go into place, you might have it in the wrong location. As you work, insert the pillar pins by hand only. Specifically, leave out the gathering pallet and the rack until the first adjustments are done.

The Pin Wheel. Refer to Figure 57 as you go on. Turn the strike train by finger pressure until the locking wheel (18) comes around and the lock pin (19) contacts the locking lever (20) to stop the gears. Observe the hammer tail (21). It should be free of the lift pins (23) on the pin wheel (24). If it is not, the hammer will always end up partly lifted when the clock finishes striking. The hammer will present a resistance to starting at the next hour or half hour. Usually a French clock assembled with this problem will give trouble.

Correcting the hammer position does not require you to separate the plates. Since the cock for the pin wheel arbor has been left off until now, you can change the position of the pin wheel without moving the locking wheel. It may take several tries before you can get the adjustment done correctly. Many French movements are marked for assembly, and our model for this chapter is no exception. The pin wheel has a punch mark which you can line up with the locking wheel pinion which has the

FIG. 57. FRENCH RACK STRIKE MOVEMENT

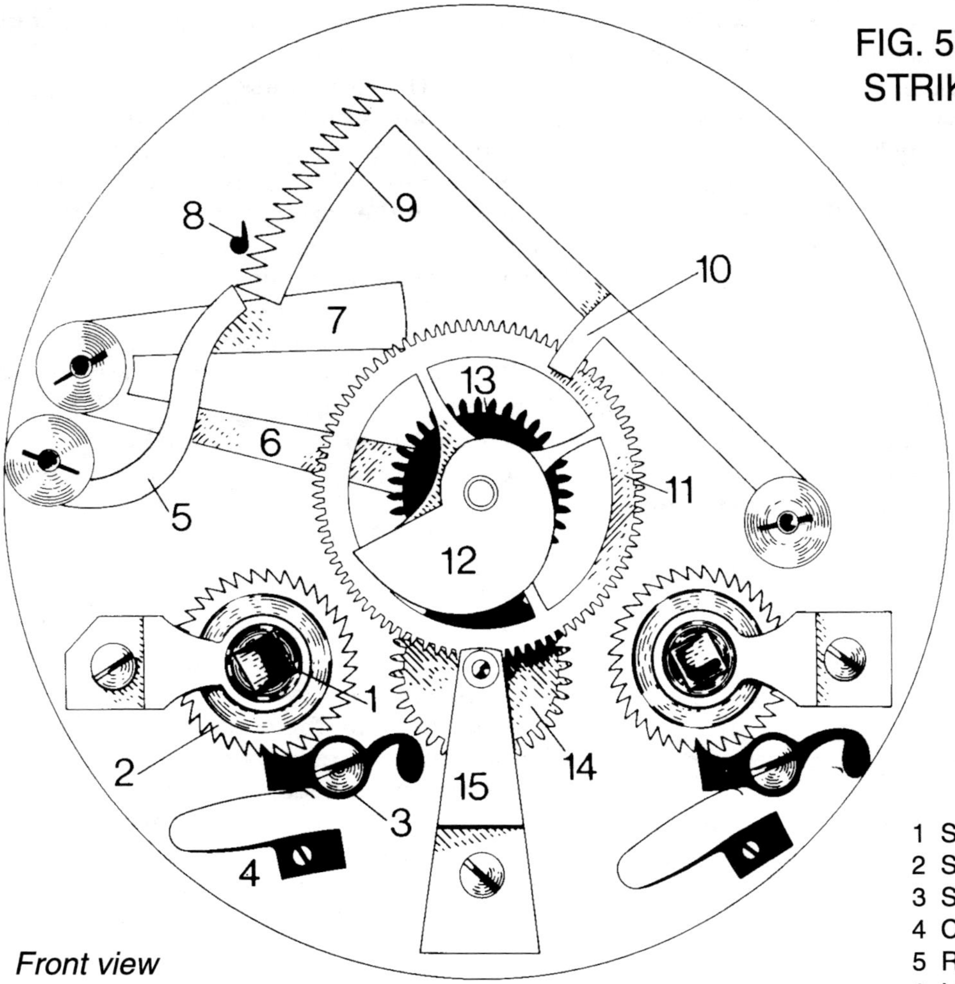

Front view

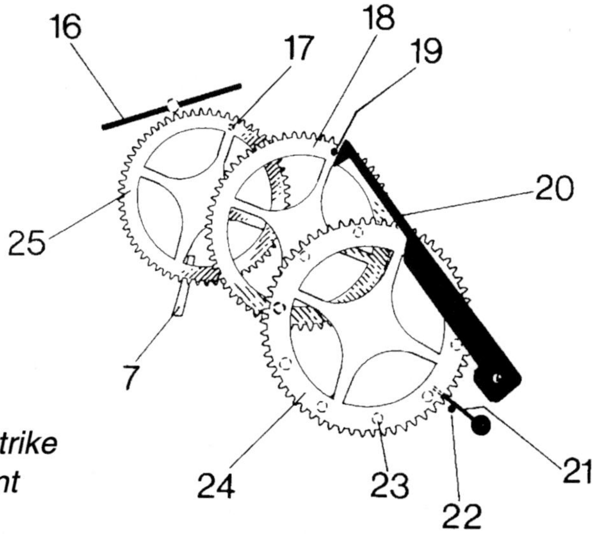

Rear view, showing strike parts inside movement

1 STRIKE WINDING ARBOR
2 STRIKE CLICK WHEEL
3 STRIKE CLICK
4 CLICK SPRING
5 RACK HOOK
6 LIFTING LEVER
7 WARNING LEVER
8 GATHERING PALLET
9 RACK
10 RACK TAIL
11 HOUR WHEEL
12 SNAIL
13 CANNON PINION
14 MINUTE WHEEL
15 MINUTE WHEEL COCK
16 FLY
17 WARNING PIN
18 LOCKING WHEEL
19 LOCK PIN
20 LOCKING LEVER
21 HAMMER TAIL
22 STOP PIN
23 HAMMER LIFT PINS (10)
24 PIN WHEEL
25 WARNING WHEEL

corner ground off one leaf. I find it easier to proceed without using the marks. Once you have completed this adjustment and have installed the cock for the pin wheel arbor, the clock should be able to strike once (since the gathering pallet and rack have not yet been added). As the strike finishes, the hammer tail must be clear of the hammer lift pins.

The Warning. The next adjustment to check is the warning pin (17). Install the lift and warning levers (6 & 7) and the rack hook (5). When the lock pin rests against the locking lever, the warning pin should be located approximately as shown in Figure 57 (bottom) which is a rear view. This position permits a half revolution of the warning wheel before the warning pin contacts the warning lever. If you need to change the warning wheel setting, first make sure there is no mainspring power on, then ease the plates slightly apart. Leave two pillar pins in place so the whole movement doesn't come apart. Move the rear warning arbor pivot out of its hole, without forcing it, and turn the wheel around to the desired position, before putting the pivot back in place. Check the strike train for correct warning and locking action before going on.

The Gathering Pallet. Next, install the gathering pallet (8) and the rack (9). The gathering pallet should be oriented so that it points away from the rack, at a 10 to 12 o'clock position (Figure 57), when the strike train is at rest. The gathering pallet rotates once with each turn of the locking wheel, and naturally it gathers up one rack tooth for each hammer blow. When the rack hook drops underneath the end of the rack, the locking levers moves down to stop the lock pin. At this point, the gathering pallet must be clear of the rack.

Finish the Assembly

Press the pillar pins in tightly because you should not have to separate the plates again. You are ready to install the cannon pinion (13), minute wheel (14), hour wheel and snail (11 & 12) and minute wheel cock (15). Adjust the snail position before installing the minute wheel cock. This adjustment is accomplished on most French clocks, including this one, by lining up marks on the hour wheel, cannon pinion, and minute wheel. When they are in line with each other, install the minute wheel cock. Double check by verifying that the rack tail (10) falls to the lowest point on the curved snail for 12 o'clock and rests on the highest part of the snail for one o'clock.

The reassembly is nearly complete now. Install the click wheels (also called ratchet wheels), ratchet cocks, clicks, and click springs. Fit the pallets and the back cock. Make sure the lifting lever and the rack move freely. The lifting lever, in particular, must not be too loose, either. This is because the cannon pinion, which has the two strike lift pins driven into it, must lift and release the lever every hour and half hour. If the lever is loose, it may drop off the pin too soon and cause erratic striking. Excessive clearance under the pin and washer can be corrected by using a very thin washer or shim under the tapered pin.

Our clock is a gong strike; there is no bell mounted on the movement. But for either gong or bell movements, the hammer must be adjusted. With the movement installed in the case, check the hammer action. It should hit the gong (or bell) once for each strike and not bounce several times upon it, or rest on it between strikes. Bouncing may be uncontrollable unless you give some attention to the pin which projects from the rear plate under the hammer arbor. Bend it slightly, if necessary, so that the hammer will contact it as the hammer hits. This will limit the bouncing. Without it, the stop pin inside the movement is the only means of restraining the hammer. Both should act upon the hammer. If you should have to bend the hammer shaft to make it closer to or further away from the gong or bell, use care. Grasp the shaft, with a pair of pliers, near the end that is away from the hammer head. Bend with another pair of pliers held closer to the hammer head. Form the hammer shaft gently; do not make sharp bends. Check to be sure the clock begins striking as exactly as possible on the hours and half hours. You can make the clock strike earlier or later by moving the minute hand collet slightly. If only the hour (or the half) is incorrect, you might be able to bend the strike lifting pin located on the cannon pinion. However, the pin is short and may not be "adjustable" in this way.

Finish up by testing the movement for a week or two inside the case. Detecting an intermittent strike defect is often more difficult with this rack and snail striking mechanism than with the locking plate type. A clock might strike only 11 times at 12 o'clock, for example, and strike all the other hours correctly. The movement readjusts itself after most counting errors, unless the defect is severe enough to cause continuous striking or no striking at all. Listen for any incorrect count, especially at 12 or 1 o'clock. After one or two successful eight-day runs, the repair is considered complete.

13

SETH THOMAS NO. 120

With round plates only 3-5/8 inches in diameter, the Seth Thomas No. 120 strike movement fits well in small arch-top and tambour mantel clocks. Figure 58 shows the No. 1 Tambour, a model 15-1/2 inches wide and 7 inches high. The No. 120 is also found in the Crandall, a 28-3/4 inch banjo clock with George Washington and Mount Vernon panels.

The rear mounted hammer strikes the hours and half hours on a coiled gong. In the tambour, the movement (Figure 59) is strap-mounted in the manner of French mantel clocks, so the movement and dial are removed as a unit through the front of the case, after the strap mounting screws are taken out. In the banjo clock, the dial comes off separately. To make the No. 120 movement easier to handle, remove the suspension unit from the back plate. Next take off the hammer bracket and the hammer arbor.

Disassembly

The two mainsprings are housed in mainspring boxes which serve only to limit the movement of the springs as they unwind and expand. They are screwed to the lower front movement plate and are stationary. The main wheels are positioned over the open back ends of the boxes, and they, of course, do turn as they power the time and strike trains. A click is located on each main wheel. These must be released and the mainsprings fully let down before any disassembly is started. A word of caution: the click springs are hard to repair if pulled away from the plane of the main wheel. Pry the clicks back carefully by lifting them with a lever. Always use a let-down key.

Fig. 58. No. 1 Tambour with No. 120 strike movement

The mainspring boxes can be taken off the movement now along with the lower front movement plate. These parts are seen in Figure 60. The rack (13) will also come away with the lower plate. Remove the two screws holding each mainspring box in place. Now hold the barrel and use a let-down key to gently turn opposite the winding direction. This should unhook the main wheel and arbor from the mainspring. Set aside the wheel, leaving the mainspring in the "box".

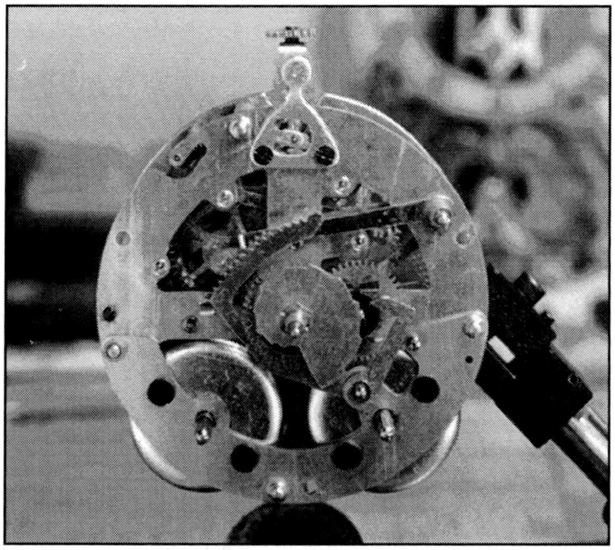

Fig. 59. Front view, Seth Thomas No.120 movement

Now you are faced with the problem of how to remove the mainspring safely without simply pulling it out with pliers, a technique which could distort the spring into a cone shape. A mainspring winder is the answer, but some winders will be difficult to use on this application. Because of the fact that the No. 120 mainsprings are housed in containers, which are open at the rear, you cannot use the clock's own arbor in the winder. The Keystone winder has three plain winding arbors of different sizes. They are like winding arbors without wheels attached. The one which fits best is hooked onto the spring in place of the rain wheel and arbor, and then the spring is removed in the usual manner.

Referring to Figure 60, remove the snail (14) and hour wheel (16). The cannon pinion (8) is held in place by a small set screw which must be backed partly out.

Removing the gathering pallet (2) requires care. The long pivot which carries the pallet is soft and will bend readily if you pry from one side. Grasping the pallet with pliers and twisting is not a good idea, because you will snap the pivot if the pallet is tight. Use two opposing levers if you must pry. Another way of removing the gathering pallet is to disassemble the movement first, then support the front plate on each side of the gathering arbor. Place a shim between the underside of the gathering pallet and the front plate, because the steel pallet pin extends through the pallet and would cause uneven pressure on it. Tap the end of the arbor, and it should drop through, leaving the gathering pallet by itself on the front plate.

Reassembly

Following the complete disassembly of the movement comes the usual cleaning, pivot, and bushing work. Part of this procedure is the inspection of two wire springs. One is the rack spring. If it is kinked and tired-looking, make a new one. Counting errors can occur if the rack does not move in a positive manner; gravity is not enough to operate the rack. The other wire spring is on the between-the-plates arbor which carries the warning lever (4), unlocking lever (5), and lifting lever (6). Check this spring and wind on a new one if necessary. Note that the locking lever (10) operates by gravity and does not need a spring.

Reassembly and adjustment of the Seth Thomas No. 120 are made easier by the fact that you can put the whole movement together including mainspring boxes, and then make your adjustments, without separating the plates again to change the mesh of wheels and pinions. Do not forget to install the spacer washer on the lower movement pillar. This spacer compensates for the overlapping of the lower front plate over the main plate. Without the spacer, you will have binding and the clock will not run properly. Add the hammer arbor, rack, and all front movement parts except the snail and minute wheel. Carefully press on the gathering pallet, leaving it slightly outward from its normal position, so that it will be easy to adjust without fear of breaking the pivot.

Adjusting the Gathering Pallet

The strike train is adjusted through the careful positioning of the gathering pallet. Moving it to exactly the right position will adjust the locking and warning action in the strike train. First check the strike operation by turning the wheels through a cycle. You will see that the rack supports the rack hook during striking, keeping the lock pin (9) away from the locking lever (10). As the rack hook (1) drops off the end of the rack, the locking lever stops the pin. Notice the position of the hammer. If the hammer lever (11) is clear of the hammer-lifting star (12), then the hammer will be at the rest position at the start of the next strike sequence. The idea is to avoid having the hammer already under load as the strike train starts up. The wheels should reach full speed before the hammer-lifting star reaches the hammer lever.

The gathering pallet makes one full turn for each hammer stroke. By moving the gathering pallet to a different place on its pivot, you can make the hammer finish up with the hammer engaged or free. By trial and error, change the position of the gathering pallet, until the strike train stops at the end of its cycle with the hammer at rest. That is half of the adjustment procedure.

The other half of the gathering pallet adjustment procedure depends on some close observation. Op-

FIG. 60. STRIKE MECHANISM OF SETH THOMAS NO. 120 MOVEMENT

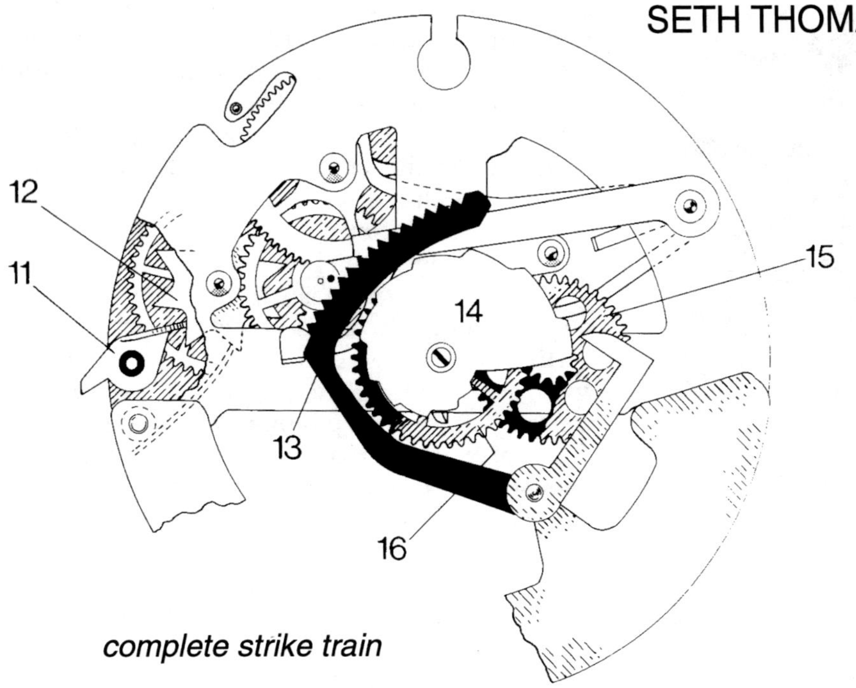

complete strike train

1 RACK HOOK
2 GATHERING PALLET
3 LOCKING WHEEL
4 WARNING LEVER
5 UNLOCKING LEVER
6 LIFTING LEVER
7 STAR CAM
8 CANNON PINION
9 LOCK PIN
10 LOCKING LEVER
11 HAMMER LEVER
12 HAMMER-LIFTING STAR
13 RACK
14 SNAIL
15 MINUTE WHEEL
16 HOUR WHEEL

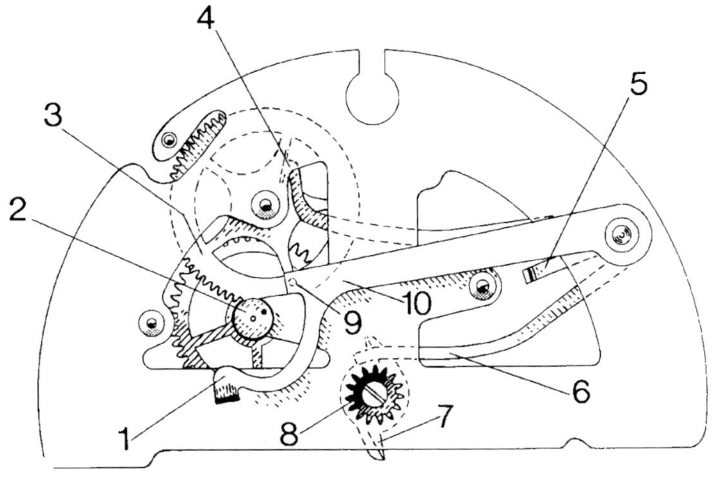

strike train partially disassembled

erate the gear train, keeping a finger near the fan or fly to stop the wheels at the right instant. Watch for the exact point at which the rack hook drops off the end of the rack. Stop the action here, without waiting for the strike train to lock itself. Check the position of the lock pin (9), which is mounted on the locking wheel (3). If the pin is at a 12 o'clock position, it is correct. It means that the locking wheel will move approximately another half turn before the lever stops the pin. This allows the gathering pallet pin to move away from the rack as shown in Figure 60. At the next strike point, the rack will be free to drop onto the snail without interference. In addition, the lock pin will not get in the way of the locking lever as it moves downward. To adjust the action of the lock pin, move the gathering pallet ever so slightly, only a few degrees, and recheck. When you have obtained the half-turn run of the locking wheel and pin, the adjustment is correct. The only difference between this adjustment and the one described earlier for the hammer is the amount of movement.

Adjusting the Hour Wheel and Snail

To finish up the strike train assembly, add the hour wheel and snail (parts 16 &14) and the minute wheel (15). Adjust the snail position before installing the washer over the minute wheel pinion. If you make sure that the rack tail falls cleanly onto the 12 o'clock and 1 o'clock positions, then all the other hours should be correct. In Figure 60, the rack and snail are in the 1 o'clock position.

Test the movement following assembly. If the strike train is not properly adjusted, you may notice that the clock fails to strike at alternating hours and half hours. This is because the gathering pallet pin is not clear of the rack at the end of the strike cycle. If the rest position of the hammer is not correct, the strike may stall or fail to work for eight days. Look at the hammer on the rear of the movement and verify that the gears reach speed before the hammer is lifted. If necessary, repeat the adjustment procedure for the gathering pallet.

14

NEW HAVEN RACK & SNAIL STRIKE

The banjo style New Haven striking clock featured in this chapter is quite different from the black New Haven count wheel striking clock in Chapter 3. Banjo clocks are perennial favorites, ranging from the valuable early weight driven examples to more recent versions with inexpensive lever movements. The New Haven clock shown in Figure 61 is not one of the earlier banjos, but it is still worthy of our attention. Probably made around 1929, it is approximately 19" tall including the brass finial at the top.

The movement is 3-1/2" in diameter with a rack strike mechanism. It strikes the hours and half hours on a single chime rod which extends from the upper waist area down to the bottom door. The front plate of the movement has three dial pillars which are inserted into the dial plate and held with the square-headed set screws favored by New Haven. The movement and dial unit is removed and inserted through the front of the case and held in place by two clamps which screw onto the dial plate. As is typical in banjo clocks, the movement is a tight fit in the case.

Any example of this clock will probably need disassembly, cleaning, and bushing of worn pivot holes. The pivots are as small in diameter as those in French clocks, so extra care is necessary when they are handled. The strip pallet recoil escapement will need the usual attention—grooves will have to be polished out of the pallet faces, the pallets "closed" if needed, and the center distance adjusted.

Fig. 61. New Haven Winetka banjo clock

Figure 62 shows the front of the movement and the rack and snail strike mechanism. The New Haven has the same basic mechanism as other rack strike movements. It's also true, however, that the shape and arrangement of some of these parts are unique. The unlocking and warning lever, for ex-

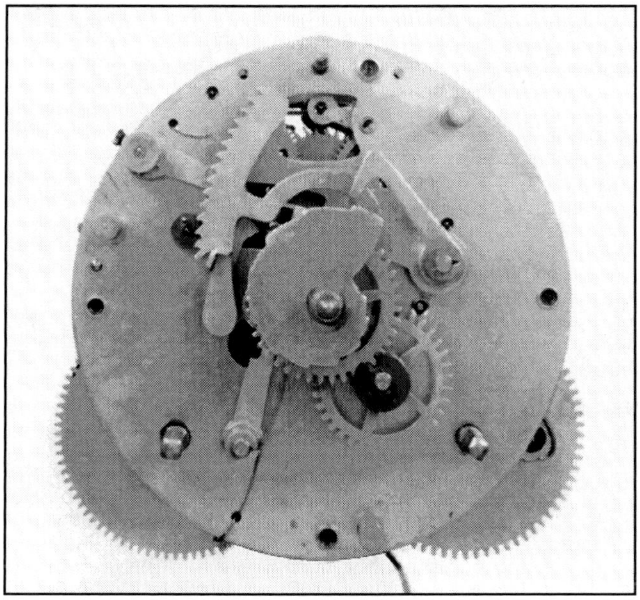

Fig. 62. Front view of the New Haven movement

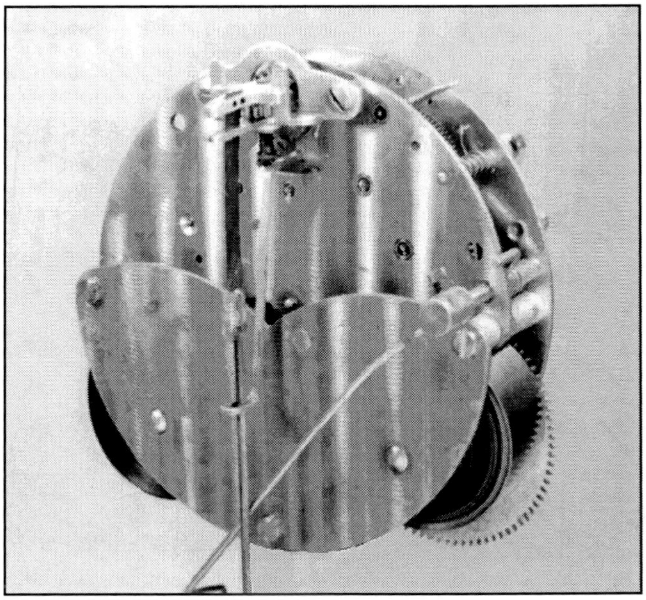

Fig. 63. Rear quarter view showing extra rear plate

ample, is pivoted from a point just to the right of the strike winding arbor. It is almost vertically oriented and is operated by a brass wire spring. The rack hook is pivoted from about a 10 o'clock position on the front of the movement, and it is mounted on the end of the arbor which carries the locking lever. The gathering pallet is the round, flat type with a single pin to gather the rack teeth as the strike notes are counted.

The rear quarter view, Figure 63, shows that the back of the movement is made up of two separate plates, with the lower plate extended outward. As a result, the main wheel assemblies are given the extra depth they need. In addition, these mainspring assemblies can be accessed without complete movement disassembly.

Mainsprings

The two mainsprings are the loop end type and are 9/16" wide. The time spring is .015" thick and about 68" long. The strike spring is only .009" thick and about 75" long.

In most movements with loop end mainsprings, it is necessary to release the springs into wire loops. These loops serve to contain the power of the springs and keep them to a manageable size. Unfortunately, the mainsprings in this New Haven movement are hooked to pillars riveted to the back plate, near the center. It would be awkward to fit wire loops or clamps over them. An easier way to control the springs is to simply release them with a let-down key. Since they are not as strong as most other American clock springs, they can be let down and expanded until the power is completely gone. You may need to turn the let-down key, rather than just allowing it to spin under spring pressure, to release all the spring power. Figure 64 shows the springs fully let down.

Double check that the main wheels are not under mainspring power, then remove the three screws which hold the lower rear plate. Carefully work the plate and the two main wheels off the movement. Now the main wheels and mainsprings can be cleaned and repaired as necessary. Don't lose the two brass spacers which fit under the lower rear plate.

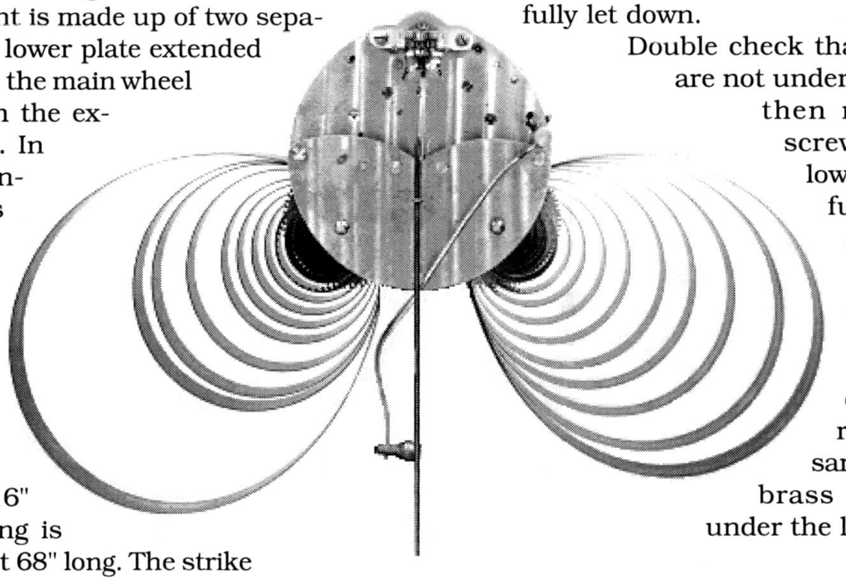

Fig. 64. The mainsprings can be completely let down, then the lower rear plate removed. This allows access to the main wheel assemblies without complete disassembly of the movement.

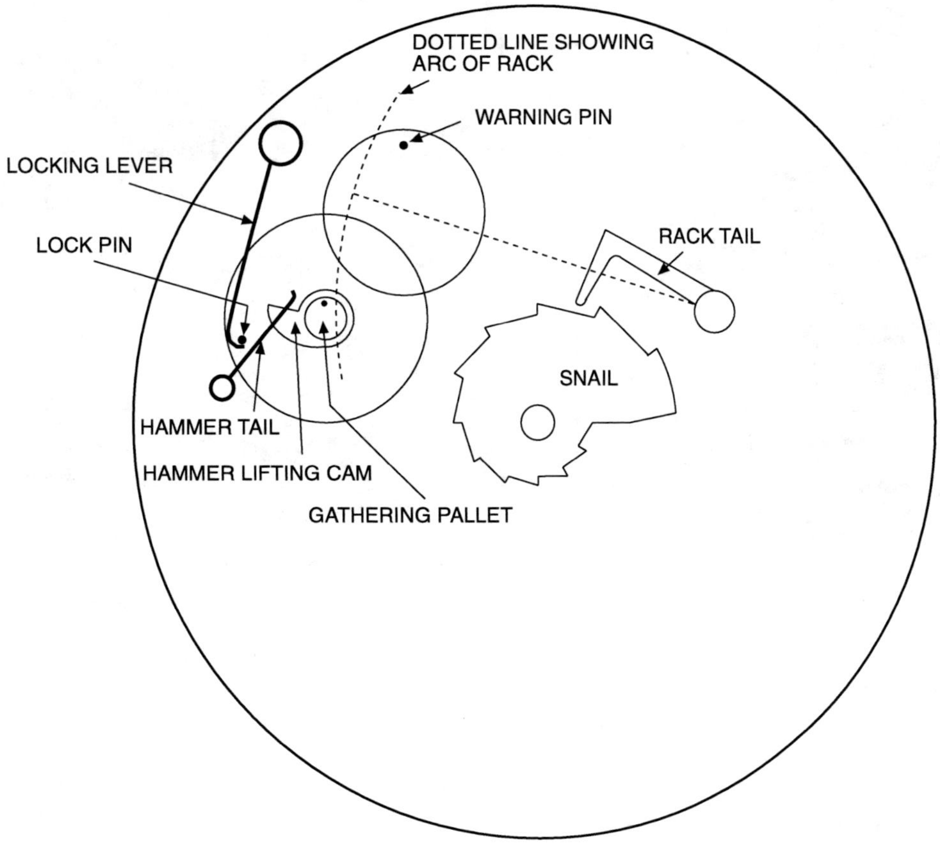

FIG. 65. SETTING UP THE NEW HAVEN RACK STRIKE

Reassembly

The reassembly process begins with the placement of the wheels and arbors into the front plate. It is not necessary to index the strike wheels in any particular way because all the strike adjustments are done after the movement is put together. Be sure to install the hammer arbor at this time. Note that it is important to wrap the brass hammer tension wire around the pillar. It would be far more difficult to do later after the movement is assembled. At this stage, the lower rear movement plate, which has the loop end mainsprings attached to its short pillars, is left aside.

Place the top rear plate over the long hammer arbor and begin to ease the rear pivots into the holes. Extra care is required for the small pivots. Tweezers will help to move pivots; avoid forcing them. When all the pivots are in the holes, install the top plate screw to hold the plates together.

Strike Adjustments

Figure 65 is a drawing of the front of the movement. A number of the parts visible in Figure 56 are not shown in order to simplify the main adjustments which are easier than they might at first appear.

In this movement, the adjustment procedure is made easier by the fact that the third strike wheel and arbor support both the hammer lifting cam and the strike lock pin. This means that you do not have to set or adjust the rest position of the hammer tail; it is automatically correct. New Haven has borrowed this characteristic from the typical American count wheel movement. It is not a feature often seen in rack and snail movements, which require a separate adjustment to keep the rack tail free of a lifting star or cam when the gear train is locked.

Warning Pin. The warning pin and wheel should be oriented in a 12 o'clock position as shown in Figure 65. This will allow almost a half revolution of warning "run". The pin moves from it's 12 o'clock position, at lock, to a 4 o'clock position, where the warning lever stops it. In this particular New Haven movement, the rear pivot of the third (locking) arbor was short and kept popping out of the hole. So, instead of handling this adjustment by easing out the rear warning arbor pivot, I used the locking arbor instead. I placed the warning pin at the desired 12 o'clock position, then moved the "unmeshed" locking wheel until the lock pin rested against the hooked locking lever. Then the third arbor was eased back into the pivot hole.

Gathering Pallet. The gathering pallet position should be set so that the pin on the pallet will be clear of the rack at the end of the strike. This assures that the rack will be free to move at the next striking point. A 12 o'clock orientation of the gathering pallet pin works well. See Figure 65.

Finishing the Adjustments. The hour and minute wheels can be added to the front of the movement. Mesh the hour wheel and the minute wheel pinion in such a way that the rack will operate correctly with the snail at 12 o'clock and 1 o'clock. This adjustment is fully explained in Chapter 10.

The main wheels and mainsprings can be added next. Before the lower rear plate is installed, place the brass spacers over the lower pillars. Finally, add the strike hammer.

Time Train Information

The time train is arranged with an auxiliary drive for the center arbor. This is similar to most American movements with count wheel strike and loop end mainsprings. The center arbor is driven off the second wheel, and it is not part of the direct gearing from main wheel to escape wheel. In contrast, most European clocks place the center arbor and center wheel directly in the gear train.

The New Haven time train gearing is counted as follows. The auxiliary drive pinion on the center arbor has 20 teeth or leaves. This is figured as a wheel, and is multiplied by the third wheel (56), fourth wheel (53), and the escape wheel (30). The total is 1, 780,800. Into this number, divide the multiple of the pinions on the third, fourth and escape arbors. These are each 8, so the number is 512. The fraction is now 1, 780,800 / 512 = 3478.125. Multiply by 2 to reach the hourly beat of 6956.25.

15

HERMLE MODEL 130-020

Modern German movements are sometimes repaired and sometimes simply replaced when servicing is needed. The difference seems to be in the repairer's attitude toward the subject. Working from a strictly monetary viewpoint, some professional repairers decide to save labor time by replacing modern movements that would need to be disassembled for cleaning and repair. The customer benefits, too, because he or she receives a brand new mechanism that will work well for years.

Other repairers, including hobbyists and some professionals, disassemble the movements, clean them, and install bushings in worn pivot holes. This may reflect a repairer's preference to restore movements instead of scrapping them, or it may be in response to a customer's desire to retain the original clock. In some cases, a repair must be done simply because a replacement movement is no longer available.

This chapter provides a strike train assembly and setup procedure for a typical modern movement. The Hermle Model 130-020 is a 3-1/2" diameter movement with a balance escapement and "bim-bam" strike on three hammers underneath the movement. It has been installed in a wide variety of cases over the years.

Figure 66 is a Seth Thomas mantel clock from the early 1960's—an old clock in modern terms! Note the styling of the case, which would have been considered up to date at the time. Reproductions of classic styles, such as bracket clocks and the ever present tambour cases, are far more common.

Fig. 66. Seth Thomas clock containing a Hermle 130-020 strike movement.

The case contains a Hermle Model 130-020 movement which is still made today. Our example, shown in Figure 67, is stamped "A206-000", the Seth Thomas number for this Hermle movement. It is also marked with the Seth Thomas date code "6113", which means the movement was manufactured during the last quarter of 1961.

It is worthwhile to look briefly at the Seth Thomas system of dating movements. In this four-digit system the first two numbers represent the year of manufacture. The last two numbers represent the quarter of the indicated year. Specifically, the first quarter was designated as -03, the second as -06, the third as -10 and the fourth as -13. Most modern clock companies used standard Hermle movement numbers (i.e. 130-020), but Seth Thomas ordered the movements stamped with Seth Thomas numbers.

Disassembly and Repair

The ratchet wheels are mounted just inside the back plate. Be sure to let down the two mainsprings before starting to disassemble the movement. Remove the crescent shaped lower front plate shown in Figure 67, then take out the two barrels. If this kind of movement is being repaired at all, it should be cleaned properly. That means the mainsprings should be removed from the barrels with a mainspring winder and inspected. Since new mainsprings are available at a reasonable cost, replace any damaged spring. Incidentally, new mainsprings should be cleaned to remove the preservative oil before they are inserted into the barrels. Lubricate the springs with mainspring grease or mainspring oil.

Separate the plates and remove all the arbors. After cleaning the parts, you should look for worn pivot holes and rough pivots. Bushings should be installed where needed. Modern clock pivots can sometimes be polished, but it may be better to replace the wheel, arbor and pinion as a unit. If the correct part is available, the cost will be small, and the new pivots may be of higher quality than, for example, those on a late 1970's arbor.

The gathering pallet must be removed from the long front pivot of the fourth arbor if a bushing must be installed at that location. The safest way to remove the gathering pallet is to disassemble the movement first. Support the front plate and the gathering pallet, then drive the pivot downward to push it off the gathering pallet. Prying from the front is not as good a procedure, even if two levers are used, since it may result in a bent pivot.

Reassembly and Adjustment

Place all the arbors into the front plate and then add the back plate. The rear pivot for the fly, along with those for two hammer arbors, is supported by a bridge on the back of the movement. There are two additional hammer arbors located at the bottom edge of the movement. These arbors are supported by their own small bridge, and they are connected to the first pair of hammer arbors by steel wires. Do not install the mainspring barrels or the balance at this stage. Screw on the pillar nuts finger tight. Add the strike lifting lever and the rack hook. If the gathering pallet was removed earlier, place it on the pivot without staking it firmly in place.

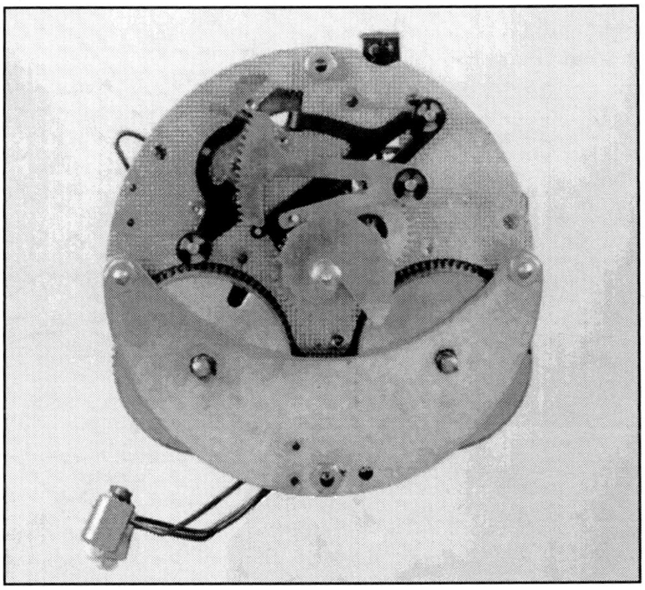

Fig. 67. Front view of the Hermle Model 130-020

Figure 68 shows the strike train parts which are most important for the adjustment of the strike train. There are several parts which need to be adjusted. These adjustments are so closely related that it is difficult to consider them separately.

Hammer Lifting Star. It is necessary to set the third arbor which carries the hammer lifting star. Arrange the wheels so that the lock pin is resting against the locking lever. When the strike train is locked in this manner, one hammer tail is seen on one side of a "point" on the star, and the other hammer tail is on the other side of the same point. Figure 68 shows the correct position of the hammer tails.

The gathering pallet should also be set as shown in the drawing. If the gathering pallet is still loose on the pivot (because it was removed during the repair of the movement) just rotate it to the correct position. If, on the other hand, the gathering pallet is tight on the arbor, it is more difficult to adjust. Separate the plates enough to ease the rear pivot of the third (hammer lifting) arbor out of its hole. Change the way the third wheel meshes with the pinion above it, on the gathering arbor. Fit the pivot in the hole again, making sure that all the other pivots are also seated. Recheck the adjustment.

More About the Gathering Pallet. The gathering pallet in this typical modern movement has two functions. The steel pin gathers rack teeth, moving the rack with each strike on the gong; the cam

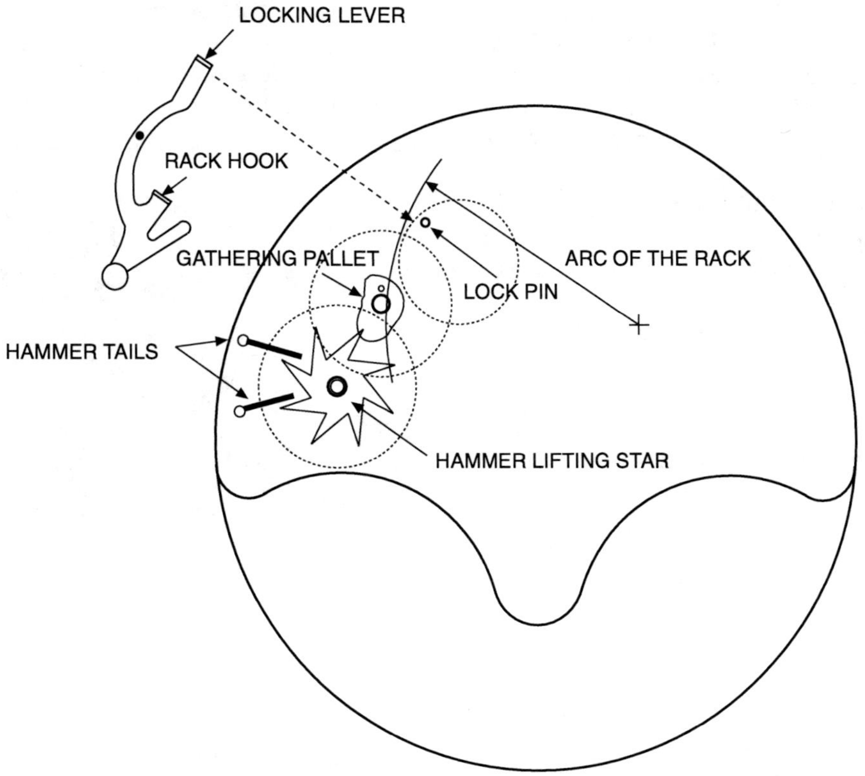

FIG. 68. SETTING UP THE HERMLE MODEL 130 STRIKE MECHANISM

shaped edge lifts the rack hook for each strike allowing the rack to move. When the strike train is locked, the pin is oriented at a 12 o'clock position. This means that the pin will not prevent the rack from moving at the next time the clock is to strike.

Whether to remove the gathering pallet during the repair is a matter of choice. If it is removed, the pallet is relatively easy to position on the gathering arbor, and this simplifies the adjustment procedure. It takes time, however, to remove the gathering pallet initially and then later stake it firmly onto the pivot again. My own choice has been to remove a staked-on gathering pallet from a modern movement only when it is necessary for pivot or bushing work on that location, although this makes the adjustment procedure somewhat more difficult. There are repairers who almost always remove the gathering pallet, and those who generally do not.

The warning. Whenever you work on a striking clock, it is important to understand how the warning action works and how it is to be adjusted. In the Hermle movement, the warning is automatically correct. The lock pin simply moves from the locking lever to the warning lever when the warning occurs. Warning occurs only about two minutes before the hour and half hour; there is usually no adjustment required, because of the flat levers which keep their shape well.

Finishing Touches

The snail should be installed and set up as explained in Chapter 10, with the strike count checked at least at the 12 and 1 o'clock positions. Check the coiled steel wire springs which operate the lifting lever and the rack hook. Each of these springs can be precisely adjusted for tension simply by turning the entire spring, which is wrapped around the hub of the lever. Tension should be strong enough to assure positive action, but not strong enough to overload the movement. Lubricate all rubbing parts of the strike train with a light clock grease. Start by greasing the star cam, which must smoothly raise the lifting lever. Both hammer tails should be greased where they are touched by the hammer lifting star. In addition, each hammer arbor has a bent wire damper which should be greased. The two sets of hammer levers, top and bottom, are connected by wires; grease the contact areas of these wires.

16

SHIP'S BELL BASICS

Instead of counting out the numbers of the hours on a gong or bell, ship's bell clocks toll the hours of the three "watches" observed on board ship. Watches are the shifts during which the sailors run the ship on a 24-hour basis. In the chapters which follow, four different ship's clocks will be covered. The greatest difference among them lies in the way each counts the hours, in particular the "odd" counts of 1, 3, 5 and 7.

The horological version of the ship's day is divided up into three equal sections for each 12-hour period. Figure 69 illustrates this idea on a clock dial. We can begin with one bell sounded at 12:30. The clock then strikes two times at the next hour (1:00) and continues to add one note to the count each half hour and hour, until the count of eight bells is reached at 4:00. One bell starts the count over again at 4:30, increasing to eight bells at 8:00. The third sequence runs in the same manner from 8:30 to 12:00. There are three sequences, called watches, in each 12-hour period.

Most ship's bell clocks are the rack and snail type. Compared to the typical snail with its gradual curve or progression of steps from 1 through 12, the ship's clock snail is divided into three segments, each with four steps. This means that 1 bell and two bells are counted on the highest "step", three and four bells on the next step, and so on. An exception is found in the Seth Thomas No. 7 ship's bell movement featured in the next chapter; it counts the ship's hours with a count wheel mechanism.

If the snail type ship's bell clock had no other special parts, it would count 2-2-4-4-6-6-8-8. For the ship's bells, however, the odd counts must also be present: 1-2-3-4-5-6-7-8. The greatest differences

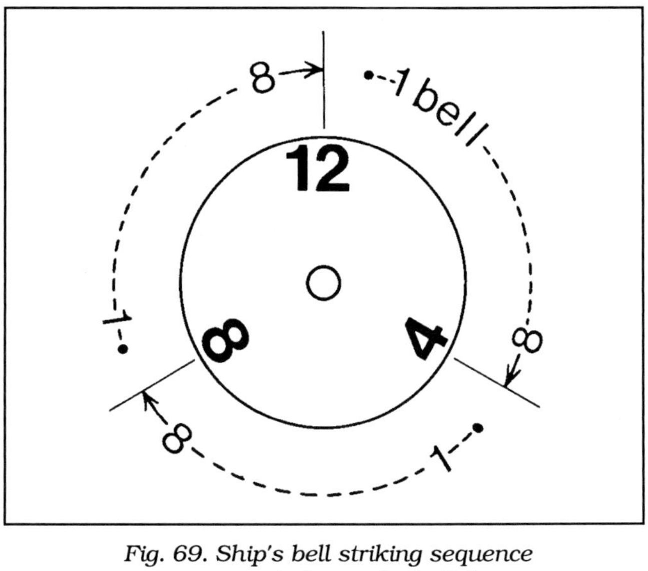

Fig. 69. Ship's bell striking sequence represented on a clock dial.

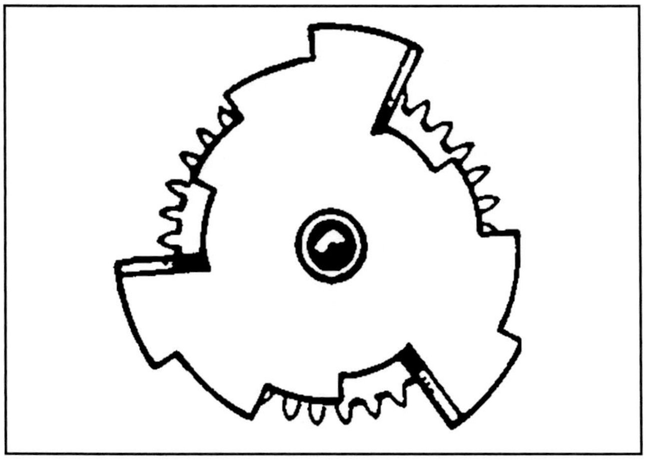

Fig. 70. A typical ship's bell snail

among ship's bell mechanisms are the ways in which the odd counts are achieved. It's fair to say that most of the assembly and adjustment problems come from this area of odd counts. Whenever you study a ship's clock, take the time to understand the way the clock achieves the odd counts. That is often the key to a successful repair.

The main counting device in a ship's bell clock is the snail, which looks like Figure 70. There are three parts to the snail, each with four steps. In most cases, the snail with four steps in each section—twelve steps in all—is used along with a device that creates the odd counts. The device generally takes the last note of an even count and holds up the hammer or suppresses the note in some way. Figure 71 shows the way this is accomplished in a modern Hermle movement (Chapter 18). The horizontal "arresting" lever snaps into place just in time to touch the hammer lever extension marked "1" in the drawing. This prevents the last note in an even sequence from sounding, thereby making it odd. The hammer still operates on that last "suppressed" note, but the lever interferes just enough to prevent it from striking the bell. Repairers sometimes have trouble with this device, which must be adjusted to work dependably every time there is an odd count.

A variation of the lever is a cam or pin which raises a lever to completely disable the hammer. It is much more positive in operation, but it still tends to cause repairers some confusion because the cam must work at the correct time.

A feature of ship's bell striking is that the bells must be sounded in pairs. That means that the note which is dropped to create an odd count must be the second note of the *last* pair. If the sequence is wrong, the strike will start off with a single note, followed by pairs of notes, then ending with a single note when the count is even. The single note should come at the end of the sequence. In the 1960–era Hermle movements, the snail had *eight* steps instead of four in each of the three segments or watches. This solved the odd count problem at the half hours, but required another complication because of the requirement that notes are sounded in pairs. The strike train was arranged to run for a silent count of "one" at about fifteen minutes before the hour. This reset the movement to begin the next hour with a pair of notes.

There seem to be more differences among ship's bell clocks than among rack striking clocks in general. For this reason, an entire section of this book is devoted to four ship's bell clocks. These are the Seth Thomas No. 7, Hermle, Schatz, and Salem (Mauthe). You will see how very different each is from the others.

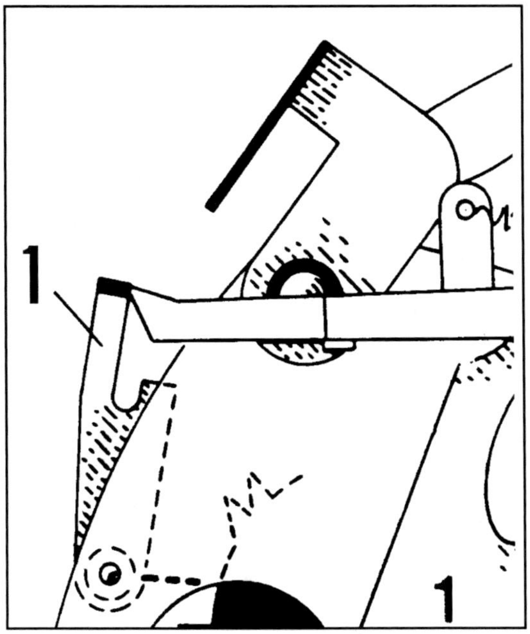

*Fig. 71. Part of a Hermle ship's bell mechanism. The horizontal "strike arresting" lever touches the hammer lever extension **1** at just the right moment to suppress the last note, making an odd count.*

17

SETH THOMAS SHIP'S BELL NO. 7

The early Seth Thomas ship's bell clocks have frustrated many repairers. These are one-day clocks which can run two or more days, although they are sometimes mistaken them for eight-day mechanisms. Figure 72 shows one model, the "Ship's Bell Lever" which features a bell mounted outside and below the clock case. Another model is the "Merrimack" which carries the bell inside the case. These early Seth Thomas ship's bell clocks are valued as collector's items. They are also a real challenge for the repairer.

Much of the repairer's difficulty stems from the fact that the movement is a count wheel type which employs two hammers to strike on a single bell. Figure 73 is a drawing which shows most of the strike parts of the No. 7 movement. Note the two hammer tails, part numbers 4 and 5, and the count wheel and count lever, part numbers 2 and 7.

Ship's bell clocks strike their notes in pairs, but on the half hour an odd count is required. Various clocks obtain this odd count in different ways. In most of them there is a means for subtracting or suppressing the last note of a pair so that the count will be odd. The old Seth Thomas movement drops the last note by utilizing 20 lift pins, 16 long and four short. The short pins can lift the upper hammer but not the lower one. Thus, when a short pin passes both hammer tails it lifts only one of them, producing the odd count.

Fig. 72. Seth Thomas ship's bell clock from catalog #531, 1892-93, courtesy of Cy Felheimer, FNAWCC

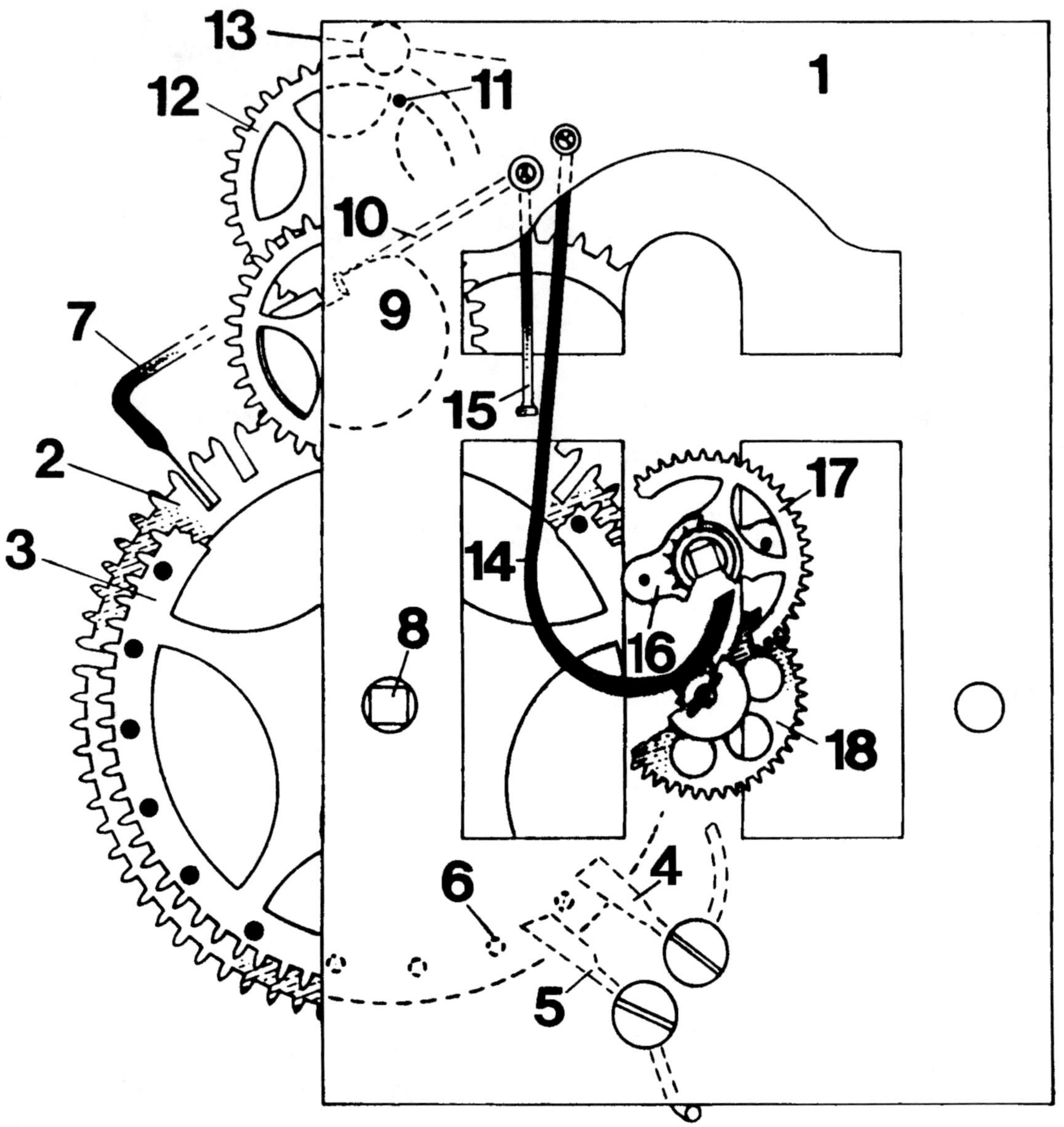

FIG. 73. SETH THOMAS NO. 7 SHIP'S BELL MOVEMENT
FRONT VIEW SHOWING STRIKE PARTS

1 FRONT PLATE
2 MAIN/COUNT WHEEL
3 PIN WHEEL
4 UPPER HAMMER
5 LOWER HAMMER
6 LIFT PINS
7 COUNT LEVER
8 STRIKE WINDING ARBOR
9 CAM
10 LOCKING LEVER
11 WARNING PIN
12 WARNING WHEEL
13 FLY
14 LIFTING LEVER
15 UNLOCKING LEVER
16 STAR CAM
17 HOUR WHEEL
18 MINUTE WHEEL

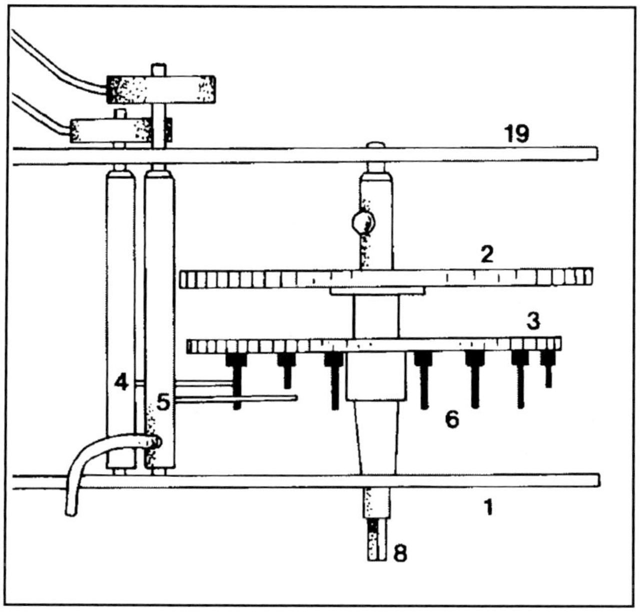

Fig. 74. Bottom view of the strike mechanism: 1) front plate; 2) main/count wheel; 3) pin wheel; 4) upper hammer; 5) lower hammer; 6) lift pins; 8) strike winding arbor; 19) rear plate.

In most count wheel movements, the position of the hammer tails does not need to be set. That is because the pin wheel and the cam are usually the very same part. There is no need to worry about whether the hammer will be left in the raised position when the clock finishes striking. In contrast, the Seth Thomas uses a separate pin wheel which must be set. Add the ship's bell striking scheme and its use of the long and short lift pins, and you have a mechanism that really throws the repairer off balance.

The pin wheel is geared to the count wheel, which explains why no amount of hand-turning or bending of levers will correct the strike sequence if the clock has been assembled incorrectly. The count wheel is marked out for a sequence of 2, 2, 4, 4, 6, 6, 8, 8, which is repeated on the other half of the wheel. The four short lift pins drop out a note each half hour to make the odd count. The result is a sequence which is 1, 2, 3, 4, 5, 6, 7, 8.

The Seth Thomas pin wheel and main/count wheel are geared together to form the ship's bell counting mechanism. As we consider this part of the movement, we can turn it upside down and look up at the two hammers, as shown in Figure 74. The upper hammer (4) is raised by each of the 20 pins on the pin wheel (3), but the lower hammer (5) is operated only by the long pins. The short pins miss the hammer tail.

Figure 75 is a diagram of a part of the train made up of two wheels and two pinions. Power passes from the main/count wheel (2) to the lantern pinion (20), and from the solid pinion (21) down to the pin wheel (3). The pin wheel is mounted on a hub (22) and would, except for the gearing, turn freely on the winding arbor.

The main wheel teeth are marked off as count wheel teeth, with deep slots at intervals to end each strike sequence. Instead of the usual hour or hour and half hour sequence, we have ship's bells as previously explained: 2 + 2 + 4 + 4 + 6 + 6 + 8 + 8 = 40. The cycle of 40 takes us through one four-hour period. Since the sequence occurs twice on the wheel, we have two periods of 40 notes for a total of 80 notes or eight hours of striking. It is necessary for the clock to drop out four notes during each four-hour period, one for each half-hour odd count. Deleting these notes is not the function of the main/count wheel, which just counts all the sequences as even: 2, 2, 4... etc. Dropping a note to make an odd sequence is a function of the pin wheel, which is the other wheel in our counting train. The pin wheel is marked off with 20 pins. This is exactly right for one four-hour period, as each pin lifts both hammers in succession. An unwanted note is dropped each time a short pin passes by the hammer tails; it can be reached only by the upper hammer, and the lower hammer is untouched. In this way, the four notes are deleted, although the movement "counts" them.

The ratio between the pin wheel and the main/count wheel is 2:1. We have seen that the pin wheel has only one four-hour period on it, while the main/count wheel has two four-hour periods. Counting

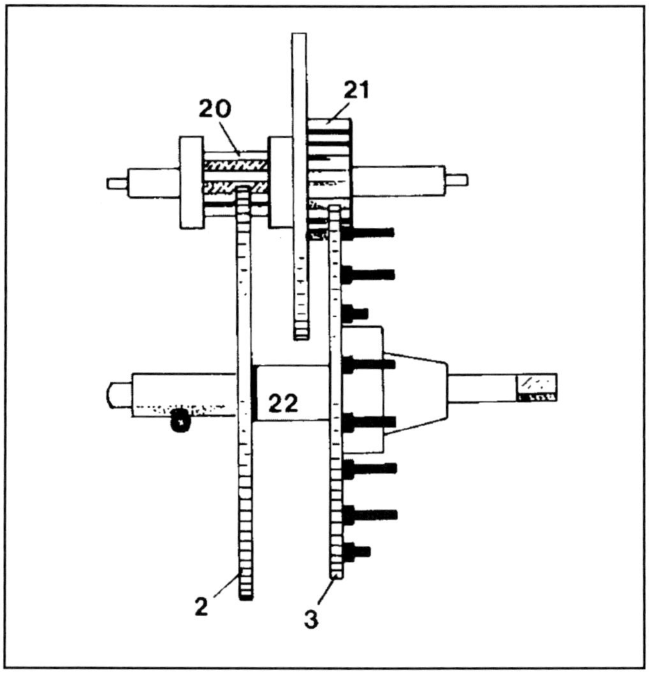

Fig. 75. Portion of the strike train: 2) main/count wheel; 3) pin wheel; 20) lantern pinion; 21) solid pinion; 22) hub.

this train shown in Figure 75, we can multiply the 80-tooth main wheel by the 16 leaf pinion to obtain 1280; divide this by 640, which is the 8-leaf pinion multiplied by the 80-tooth pin wheel. The answer is 2 (2:1 ratio).

Finally, consider that the main/count wheel has 16 deep slots. Unlocking the count lever 16 times in a clock which strikes the hour and half hour will consume eight hours of "counting" time. Therefore, the main/count wheel must turn three times in each 24-hour period, and the pin wheel will turn six times.

Another part of the strike train performs the warning and locking functions. The locking takes place on the strike cam, which has one slot. There is no separate lock pin as seen on many of the American striking clocks. The Seth Thomas ship's clock uses the locking method employed on many of the New Haven movements and on 30-hour clocks in general. The warning action must be set during assembly to provide one half revolution (or slightly less) of the warning wheel as the warning occurs. Refer back to Figure 73 (page 70) to identify the parts.

Each time the locking lever passes over the slot in the cam, the count lever dips between two teeth on the main/count wheel. When the lever drops into a deep slot, the train locks. Check the brass tension springs on the count/locking lever and the lifting/warning lever. Replace the springs if they are kinked or broken.

Assembly Procedure for the Ship's Bell Clock

We will begin from the point where the movement is completely apart and ready to be reassembled. Pick up the pin wheel by itself and hold it up in front of you with the pins facing down. Using Figure 76, find the short pin a) which has the exact pattern of long and short pins surrounding it. Now move to the long pin b) and place a pencil mark on the wheel at this point.

Next, perform the same kind of operation on the count wheel. Place a pencil mark at the base of the deep slot shown in Figure 77. There are two such slots in the wheel, opposite each other, Mark either one of the slots.

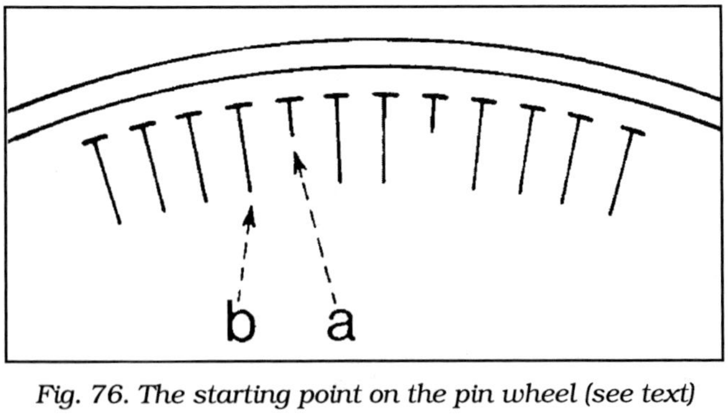

Fig. 76. The starting point on the pin wheel (see text)

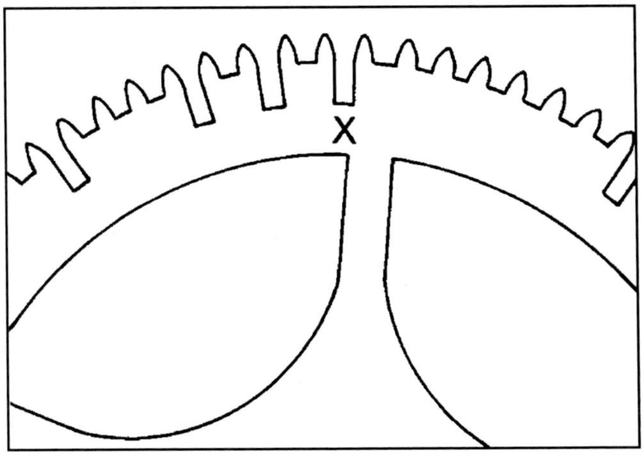

Fig. 77. Marking the beginning of the strike sequence

Assemble the movement with the count lever in the marked slot and with the marked lift pin set between the two strike hammer tails. Figure 79 (opposite page) shows how the between-the-plates levers fit in relation to one another. The strike train is to be set in the locked position. The pivots have a maddening habit of popping out during assembly as you re-index the wheels.

A Difference in the "Merrimack" Model

Another version of the Seth Thomas ship's bell movement was found in the "Merrimack" model, which has the bell mounted inside the case rather than on the outside. It appears to be a more recent model than the outside-bell clock pictured in Figure 72, and it has a useful assembly aid built into it. There is a bridge, shown in Figure 78, which supports the front pivot of the second arbor. When the mainsprings are kept fully let down, you can remove the screws and take off the bridge. This allows you to disengage the arbor without having to pull apart the plates. This makes the assembly procedure much easier. All you have to do is to put the movement together so the count lever, locking lever, and warning wheel are properly set. The count wheel should be turned so the lever is located at the beginning of the strike sequence as described earlier. Then remove the bridge and move the pivot upward to disengage it from the pin wheel. Now move the pin wheel until you reach the starting point which has also been described. Finally, replace the bridge. This

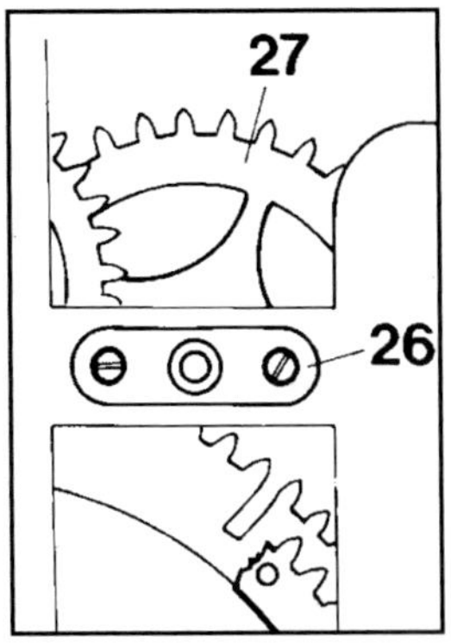

Fig. 78. Bridge 26) for the front pivot of the second wheel 27) eases assembly work

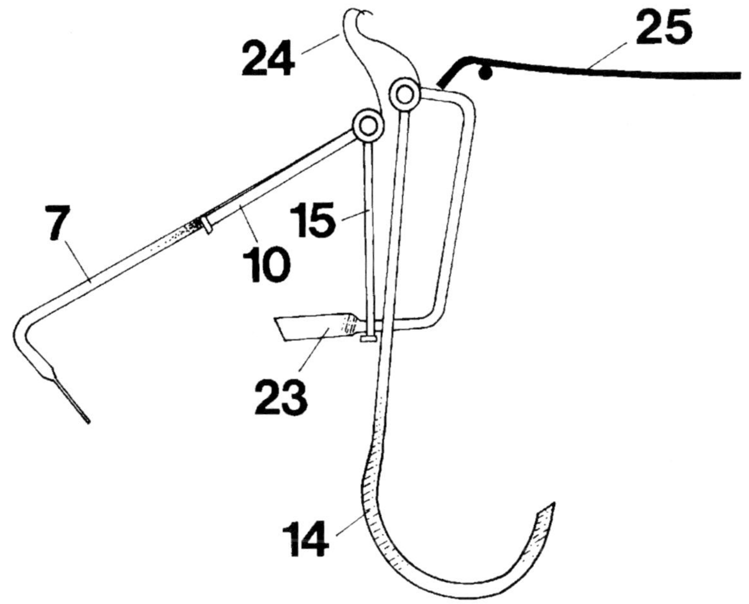

FIG. 79. STRIKE LEVERS, SETH THOMAS SHIP'S BELL CLOCK

7 COUNT LEVER
10 LOCKING LEVER
14 LIFTING LEVER
15 UNLOCKING LEVER
23 WARNING LEVER
24 WIRE SPRINGS
25 STOP SPRING

method eliminates much of the frustration caused by pivots popping out of place. There is less removal and reassembly of parts involved. Perhaps Seth Thomas designers added this feature to speed up assembly at the factory. As repairers, we can also appreciate the improvement.

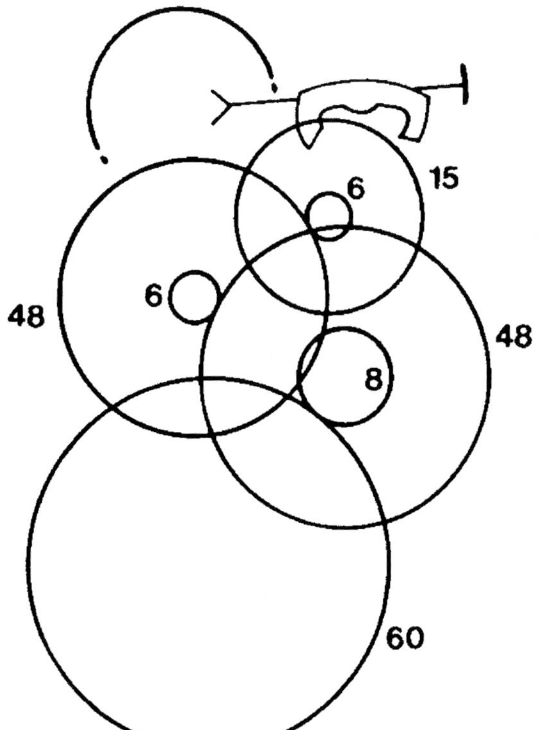

Fig. 80. The time train of the Seth Thomas No. 7 movement beats 14,400 times per hour. The numbers on the drawing are tooth counts (see text).

Movement Data

It will be helpful to mention that the mainsprings I have used for the Seth Thomas No. 7 ship's bell movement are: time, 5/8" x .018" x 96" hole end; and strike, 5/16" x .016" x 72" loop end. These sizes are mentioned in the book, *The Best of J.E. Coleman: Clockmaker,* always a valuable source of information.

The time train is balance controlled and beats at 14,400 per hour. "Counting" the train to obtain this figure is done as follows, referring to Figure 80. The wheels are 60, 48, 48, and 15, working from the center wheel up to the escape wheel. The pinion counts are 8, 6, and 6. Multiplying the pinions together and diving this number into the multiple of the wheel tooth counts yields 7,200. This is doubled because of the fact that each escape wheel tooth is touched twice (once by each pallet). The result is 14,400 beats per hour, a common train count for old lever movements.

18

HERMLE SHIP'S BELL MOVEMENT NO. 132-071

The Hermle ship's bell movement has been used by Howard Miller Clock Company, Seth Thomas, and other modern clock manufacturers. This balance controlled striking clock movement is similar to the Hermle Model 130-020 featured in Chapter 15. The important difference is that the Model 132-071 ship's bell movement strikes the bell sequences of one through eight bells instead of the hour and half hour count.

Figure 81 shows the front of the ship's bell movement. The two mainsprings are contained in barrels. After you let down the mainsprings, you can take out the barrels by removing the crescent shaped plate on the front of the movement. The clicks and click springs are mounted inside the rear movement plate. The ratchet wheels fit on the squared rear portions of the winding arbors. It is important to avoid mixing up the two barrels. The time barrel is marked "30" on the cover; the strike barrel is marked "31". As you remove the mainsprings from the barrels for cleaning, mark one of the ends so you can easily identify it later. The strike mainspring is slightly thicker and 50mm (about 2 inches) longer than the time mainspring. Don't skip the cleaning of the mainsprings. If they are dirty and sticky, they will not deliver smooth power to the gears.

Fig. 81. Hermle Ship's Bell Movement No.132-071

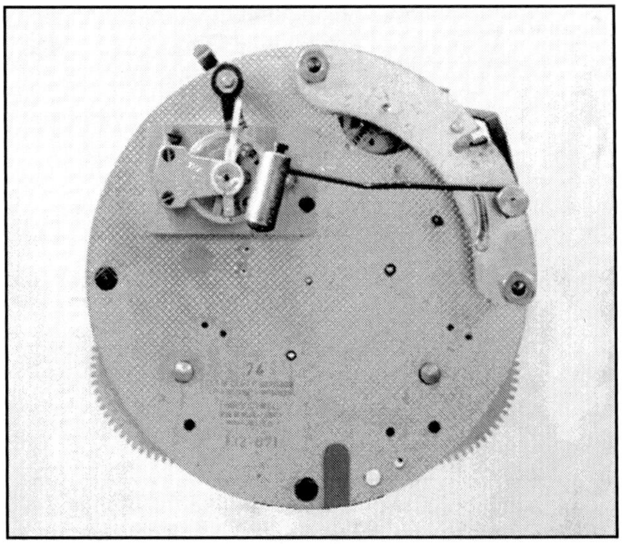

Fig. 82. Rear view with the bell removed

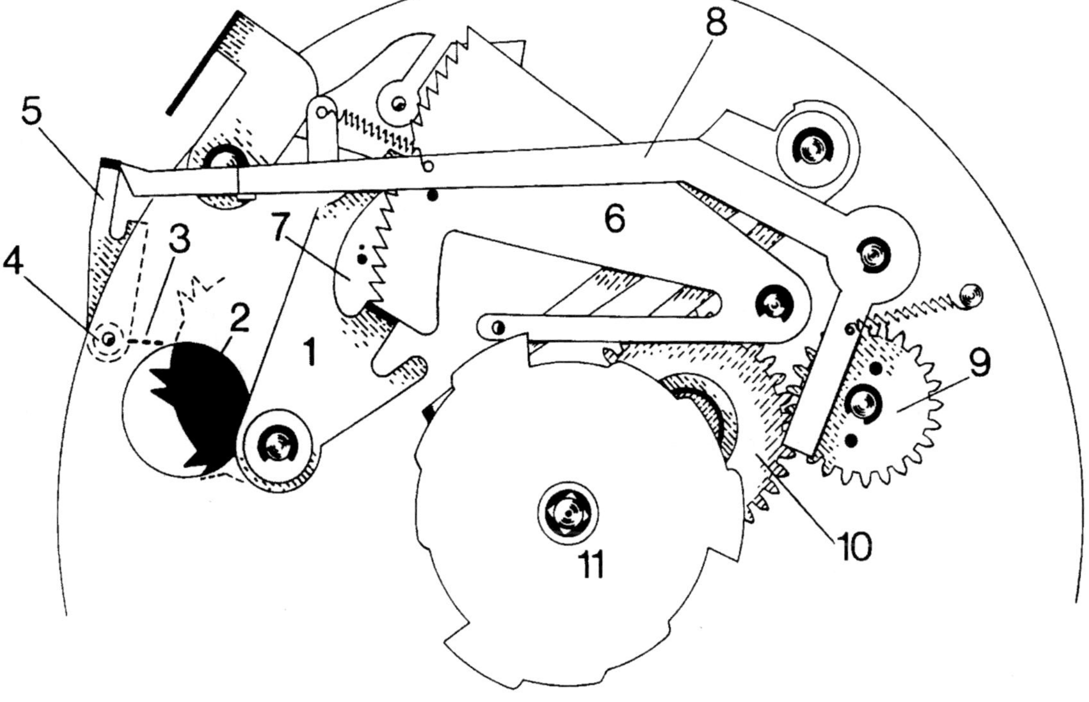

FIG. 83. FRONT VIEW OF THE STRIKE TRAIN AFTER THE *HALF HOUR* STRIKE

1 RACK HOOK
2 HAMMER LIFTING STAR
3 HAMMER TAIL
4 HAMMER ARBOR
5 HAMMER LEVER EXTENSION
6 RACK
7 GATHERING PALLET
8 STRIKE ARRESTING LEVER
9 LIFT WHEEL
10 MINUTE WHEEL
11 SNAIL

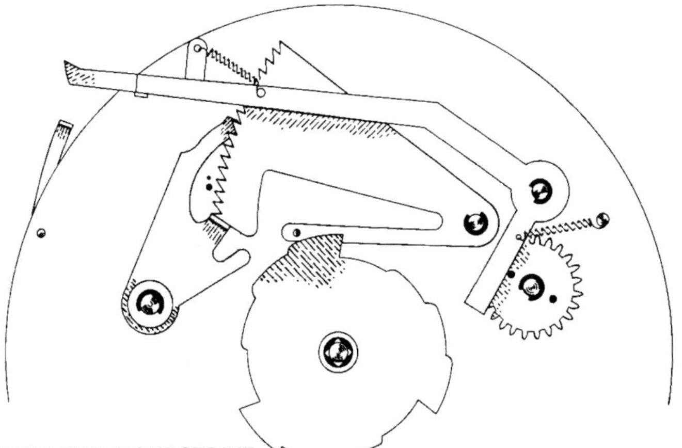

AFTER THE *HOUR* STRIKE

Figure 82 shows the back of the movement with the bell removed. The platform escapement is visible at the top left side of the plate. A single hammer is used in this movement.

Understanding the Hermle Ship's Bell Strike

Before starting any repair work on this model, take time to understand it first. Figures 83 and 84 identify the various parts. As a basic rack and snail mechanism, the Hermle has a star cam (13) with two equal arms. This cam produces the same amount of lift at the hour and half hour. The rack is fully released each time, unlike many rack movements which produce only a partial release of the rack on the half hour. The Hermle lifting lever (17), warning lever (16), and unlocking lever (12), are all on the same piece. The rack hook (1) and locking lever (14) are on another piece. In this movement the warning run is already established, and it does not have to be set by the repairer. The lock pin (15) runs from the locking lever to the warning lever, which are about 90 degrees apart.

Other parts of the mechanism relate to the counting of ship's bells. The snail (11) has three sectors of four steps each. As explained in Chapter 16, this type of snail produces only the even counts of 2, 2, 4, 4, 6, 6, 8, and 8. On the odd hours the last hammer moves, but it does not reach the bell.

The strike arresting lever (8) is the key to the counting of odd bells for the half hours. Figure 83 shows the position of the strike train parts after the clock has completed the half hour strike. The tip of the strike arresting lever rests against the hammer lever extension (5). The hammer lever extension is connected to the hammer arbor (4), so the hammer is raised just enough to prevent it from hitting the bell. The mechanism must come into play at the right instant so that only the last hammer stroke is stopped in this way. Note the pin on the rack (6). This pin is in position to raise the strike arresting lever along with the rack, as the teeth are counted.

As the minute hand moves up toward the hour, the lift wheel (9) performs its function. The wheel rotates once every two hours; the two pins on the wheel alternate in lifting the strike arresting lever each hour. When the lifting action begins, the tip of the lever moves up, releasing the hammer lever extension. This frees the hammer, enabling it to work on the hour. Lifting continues until the strike arresting lever is completely out of the way. Then, on the hour, the full even count of bells is sounded. The last note is not suppressed.

The inset view in Figure 83 shows the strike train after the hour has struck. The strike arresting lever is at its high point, clear of the hammer lever extension. Gradually the lift wheel continues to turn, lowering the strike arresting lever until the tip rests upon the hammer lever extension. With the first stroke on the half hour, the hammer lever moves away, allowing the strike arresting lever to drop further, until it contacts the pin on the rack. Again the strike arresting lever travels up with the rack, stopping the last hammer stroke for an odd count.

The strike arresting lever is pivoted near its end and is spring loaded. The tip of the lever can be moved downward, and it returns by spring tension. If someone turns the minute hand rapidly from the half hour up to the next hour during striking, the lever would be forced against the hammer lever extension, bending both parts. But the pivoted arrangement prevents this from happening. The strike arresting lever pivots and moves past the extension without bending anything. Check the tension spring for damage because it is essential for strike train operation.

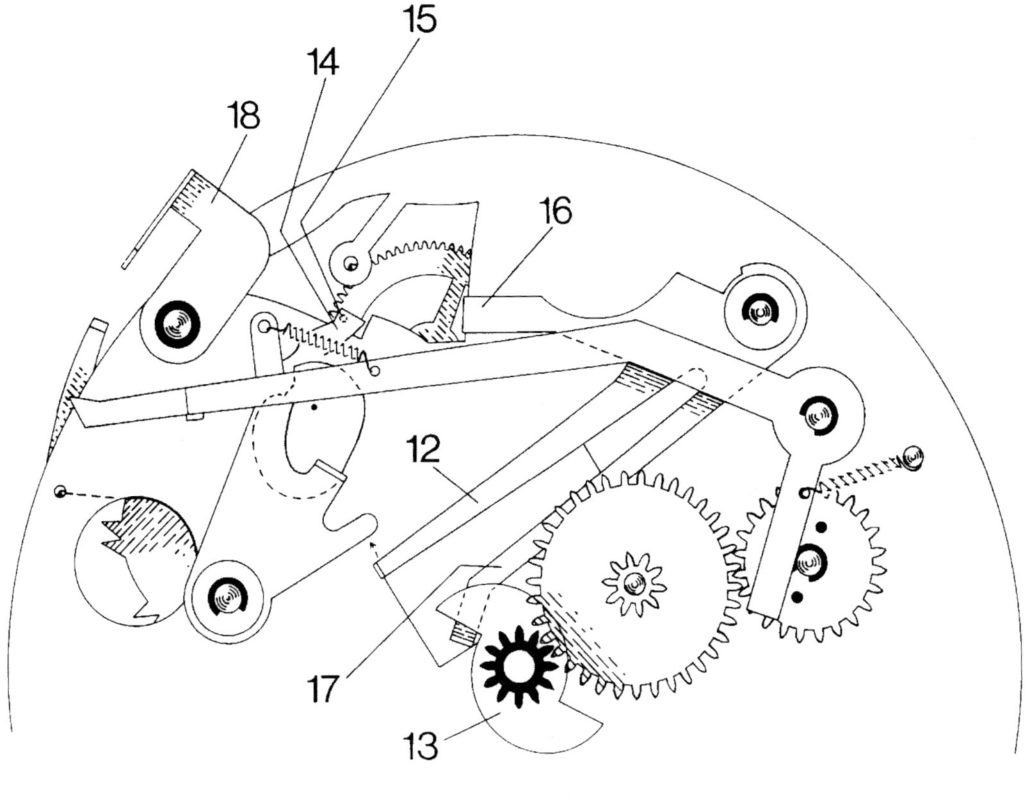

FIG. 84. HERMLE SHIP'S BELL MOVEMENT (FRONT) WITH MOTION WORK REMOVED

12 UNLOCKING LEVER
13 STAR CAM
14 LOCKING LEVER
15 LOCK PIN
16 WARNING LEVER
17 LIFTING LEVER
18 STRIKE SILENCER LEVER

Reassembly and Adjustment

After cleaning, begin the reassembly. Put all the arbors into the front plate. The mainspring barrels should be left aside for now. Get the front and rear plates together with the pillar nuts finger tight only.

Figure 85 shows the strike train parts on the back of the movement. Install the hammer bridge (21), which supports the fly (19), the strike silencer (20) and the hammer (22). Don't forget to add the two brass spacers under the hammer bridge. Adjust the strike silencer to stop the fly when the strike silencer lever (Figure 84) is moved to the "silent" position. (A final adjustment can be confirmed when the dial is in place, later on.)

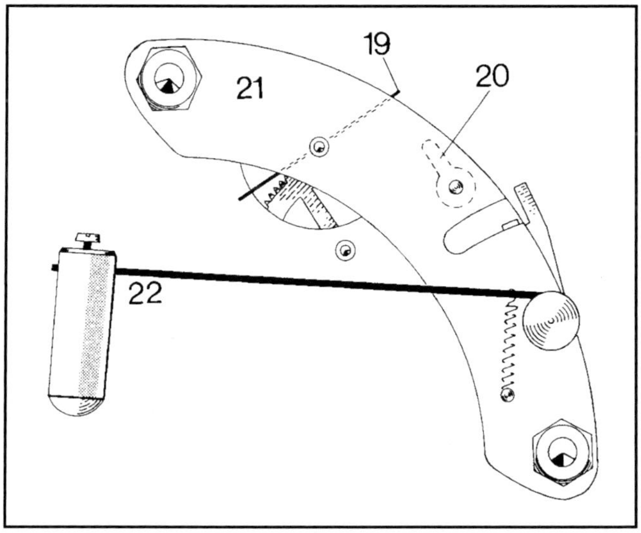

Fig. 85. Rear strike train parts: 19) fly; 20) strike silencer; 21) hammer bridge; and 22) hammer.

Refer also to Figures 83 and 84. Returning to the front of the movement, put on the lifting lever and the rack hook. Each has a wire spring which must be adjusted for tension. The entire spring can be rotated on its post, allowing you to set the tension. There should be positive tension, but not the maximum that can be obtained. You can add the rack now as you begin to test and adjust the strike train, but you will have to remove it later to get the minute wheel (10) in place.

Using finger pressure, turn the strike train gears. Now observe a few points to set up the strike train correctly. The ship's bell strike notes must be sounded in pairs, so the arms of the hammer lifting star (2) are arranged the same way, in pairs. With the strike train locked at the end of a cycle, the hammer tail (3) must rest in the wide interval between these pairs. To adjust, separate the plates enough to ease the fourth arbor rear pivot out of its hole. Turn the hammer lifting star to the correct setting. Reassemble the plates, taking care not to force the pivots into position. Check to be sure that the lock pin (15) rests against the locking lever (14), and the pin on the rack hook (1) is in the notched portion of the gathering pallet. If the parts are set this way, the gathering pallet pin will be clear of the rack teeth, permitting the rack to fall upon the snail before each hour and half hour.

Add the mainspring barrels now. The click wheels fit on the rear portions of the winding arbors. Install the lower crescent shaped plate over the arbors. The plate looks like it would fit two ways, but only one way will permit the movement mounting plate to be added later. There is a hole in the plate for the lower mounting pillar, and the hole will be located on the wrong side of the movement if you put the plate on the wrong way.

Continue with the assembly by moving the minute hand to a strike point. Install the hour wheel and snail unit as shown in Figure 83. The rack tail should fall upon one of the steps of the snail, toward the right side of the step. This will be a half hour point. Put on the lift wheel (9), meshing it with the minute wheel (10). The pins on the lift wheel should be oriented as in Figure 83 (main view). They should not be in contact with the strike arresting lever.

Tighten the pillar nuts and wind the mainsprings. Add the bell to the rear of the movement and verify that the bell count proceeds as it should. Pay particular attention to the odd (half) hours, because the last note must not be sounded when the count is odd.

Earlier Version of the Hermle Ship's Bell

When the Hermle ship's bell movement was first produced, it had a snail with three segments, each with eight steps on it instead of four. This made it unnecessary to have a "subtractor" type mechanism to hook onto the hammer and prevent it from hitting the bell at the end of the half hour strike.

Unfortunately, this setup introduced its own complication. Since the gear train actually ran for all the even and odd counts, it would finish up each odd count having struck only the first note of a pair. That would mean that the next hour would start out with a single note, not a pair of notes.

The mechanism had a way to "even out" the count. An extra (third) tab on the star cam tripped the gear train to run for one note at about fifteen minutes before the next hour. This extra note was not heard because a cam acted to hold up the hammer. In this manner, the strike train was reset to start the hour off correctly with a pair of notes.

19

SCHATZ SHIP'S BELL

August Schatz & Sohne of Germany produced a four jewel ship's bell movement which has proved to be a challenge for repairers. Figure 86 shows a seven-inch diameter brass cased clock containing the Schatz movement. The dial is marked "Made in West Germany Expressly for Swift & Anderson, Inc., Boston 25, Mass." The figures on the dial show the full 24-hour day, even indicating the names for each of the watches: afternoon watch; first dog watch and second dog watch (each only two hours in length); evening watch; midwatch; morning watch; and forenoon watch.

Ship's bells are struck on a large brass bell mounted on the rear of the movement. There are two ways to silence the strike in this clock. A selector pierces the dial at "3", raising the lifting lever to silence the clock. It sends the strike train to "warning" and keeps it there. When the selector lever is lowered, the clock resumes striking. The other method of silencing is through a screw on the case back. It is connected to a lever mounted inside the case. The lever has a notch for each hammer. When the screw is turned, the lever simply clamps down over the hammers, stalling the gear train at the beginning of the next strike cycle.

Figure 87 shows the front of the movement. Note the black finished ratchet covers. If the mainsprings are let down, these covers can be removed; then the barrels can be taken out of the clock without complete disassembly of the plates. A single click

Fig. 86. A four jewel Schatz ship's bell striking clock. The dial is marked "Swift & Anderson".

spring operates both clicks. The hooked lever at the left side is the strike detent. It catches the hammer lever extension to suppress the last hammer blow in a series, thereby creating an odd count for a half hour. This mechanism is explained later in this chapter.

Figure 88 is a rear view of the movement. The large bell has been removed in order to reveal the back plate. Note the two hammers which strike the

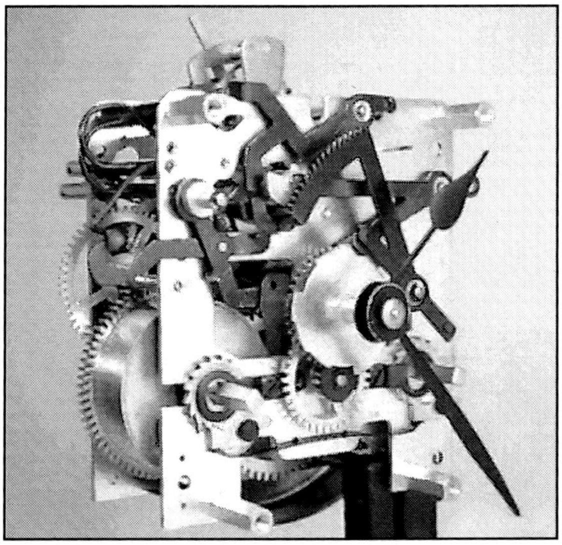

Fig. 87. Front quarter view of the Schatz movement

same bell. Also note the platform escapement, which is visible at the top of the movement. This escapement is regulated by an arbor which passes through the dial at "12".

Rack and Snail Strike Operation

The Schatz strike mechanism is the rack and snail type. Figures 89 through 92 identify the major strike train parts. The gathering pallet (10) is fitted with a pin to move the teeth on the rack (14). The outer edge of the gathering pallet is shaped as a cam which raises the rack hook (12) as each tooth is counted off. The rack hook rests between two rack teeth at each strike, holding the rack motionless until the pin on the gathering pallet comes around again to push the rack one tooth space ahead.

Strike locking is by means of the lock pin (8), which comes to rest against the locking lever (9). The lock pin also doubles as the warning pin. As the strike train moves from the rest position to the warning, the lock pin and locking wheel (7) simply move the fixed distance from the locking lever to the warning lever (11). This means that there is no "warning run" to set up when you assemble the clock. It is already established at somewhat less than a half revolution of the locking wheel, based on the location of the levers. The star cam (18) with two equal arms raises the lifting lever (15). This lever has the warning lever at one extreme, and the unlocking lever (13) at the top. Notice that the rack hook and the locking lever are made from one piece.

The third wheel carries the hammer lifting star (27). The star has ten equally spaced arms. As the clock strikes, the two hammers deliver their blows in distinct pairs. The spacing of the notes does not come from the spacing of the ten arms. It comes rather from the placement of the upper and lower hammer tails (26) and (25). These are close together, so that each arm on the hammer lifting star raises the two hammer tails in quick succession. This creates the pairing of the notes.

The clock is designed to sound eight bells at 12, 4, and 8 o'clock. From each of these three points, the sequence continues with one bell at the following half hour, two on the next hour, and so on until the count of eight bells is reached again. The hours always have an even count, and the half hours are odd. If the count is odd, the odd (not paired) note is sounded last. Five bells, for example, is sounded as one pair of notes, a pause, a second pair of notes, another pause, and a single note at the end.

The snail (16) has three "high spots" with three steps leading up to each. From this, we can visualize the sequence of 1, 2, 3, 4 which is performed three times. This might appear to give us a maximum of only four bells instead of eight. The movement does, however, give the full eight bell count. Each step on the snail represents an even count of two, four, six, or eight. The rack tail hits a step at the hour, and then touches the other end of the same step for the following half hour. This means that the rack and snail mechanism counts 2, 2, 4, 4, 6, 6, 8, 8.

Striking the Half Hours

Counting the odd (half) hours is the key to understanding any ship's bell clock. It is the feature which really distinguishes one ship's bell strike mechanism from another. Unless you understand this aspect, you cannot repair the strike. What the Schatz does is to simply arrest the last hammer blow at each half hour. The gear train runs, counting out pairs of hammer blows each time. At the last strike,

Fig. 88. Rear view of the Schatz movement

FIG. 89. SCHATZ SHIP'S BELL MOVEMENT, FRONT VIEW AFTER STRIKING SEVEN BELLS.

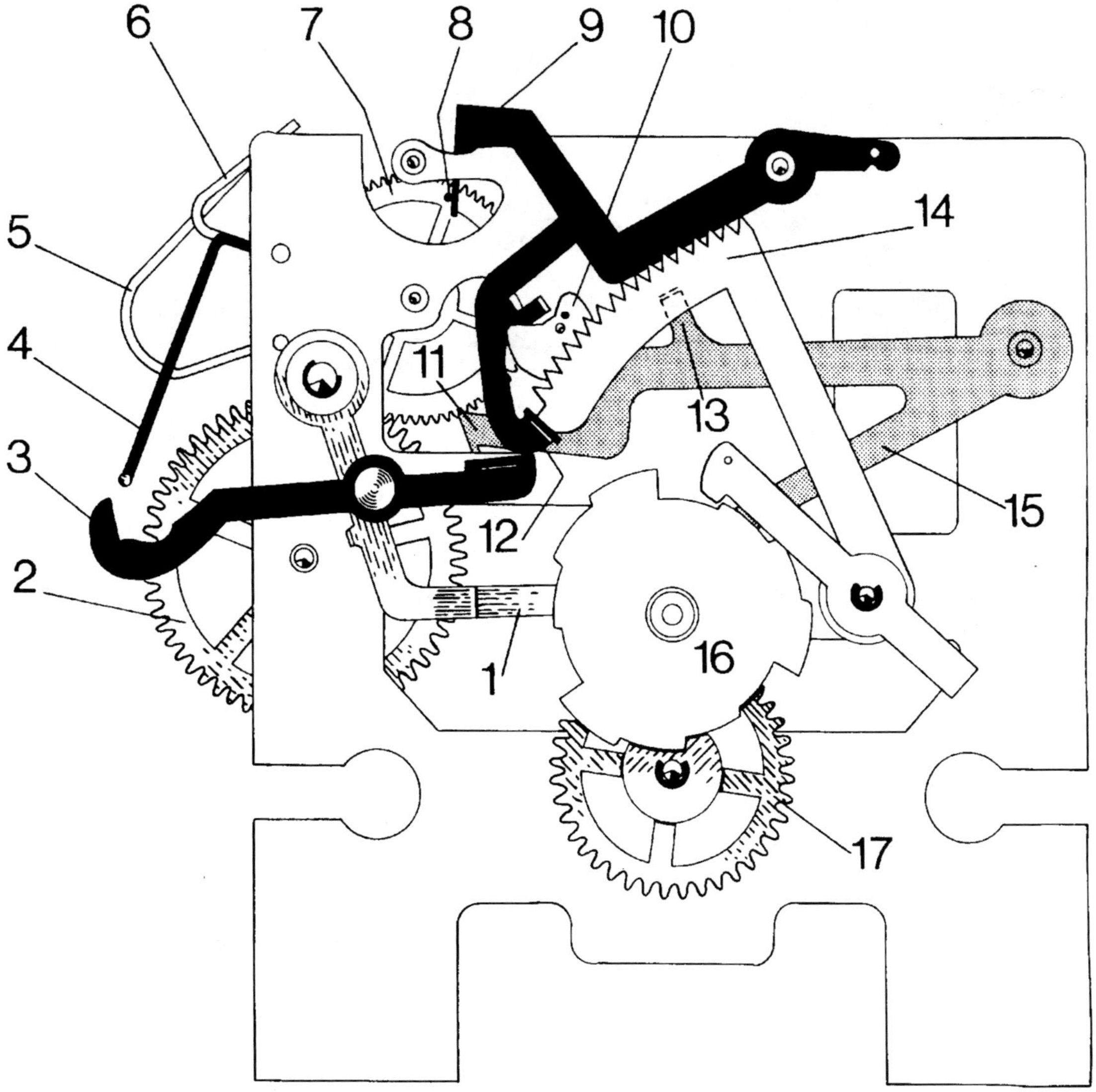

1 STRIKE ARRESTING LEVER
2 STRIKE SECOND WHEEL
3 STRIKE DETENT
4 HAMMER LEVER EXTENSION
5 LOWER HAMMER LEVER
6 UPPER HAMMER LEVER
7 LOCKING WHEEL
8 LOCK PIN
9 LOCKING LEVER
10 GATHERING PALLET
11 WARNING LEVER
12 RACK HOOK
13 UNLOCKING LEVER
14 RACK
15 LIFTING LEVER
16 SNAIL
17 HOUR WHEEL

the hammer moves but just doesn't reach the bell. There is no sound at this last stroke. The mechanism for this is shown in Figure 91.

Here's how the mechanism works to count the odd (half) hours. The strike arresting cam (20) comes into play as the half hour approaches and an odd count is always required. The strike arresting lever (1) rides up on the cam, which pulls the lever upward and to the right. This movement positions the strike detent (3) so it can do its work. The right end of the detent touches the rack hook, moving with it as the clock strikes. At the end of the strike sequence, the rack hook drops to its low point at the end of the rack. Pushing the detent down at this end moves its opposite end upward. The detent is pivoted in the middle (notice the round head of the rivet in Figure 91). As the last strike note is to be sounded, the hooked end of the detent comes up to catch and hold the hammer lever extension (4). It swings up so quickly that it is easy to miss the action. It is also easy to miss because the strike detent moves *each time* a pair of notes is struck. However, only at the last note on the half hour does the detent actually catch the hammer lever extension to prevent the hammer from hitting the bell. In this way, the detent prevents the upper hammer from sounding the last (even) note, thus creating the odd count that is needed. The hammer remains hooked until the next hour strike begins; the hammer is released just in time to operate.

At the hour, the same motions occur, but the last note is not suppressed. The strike arresting cam does not move the strike arresting lever far enough to the right and upward. The detent moves but misses the hammer lever extension. The last note is sounded, and the strike count is even.

Assembly and Adjustment

When you assemble the movement, take care to observe a few points. Referring to Figure 89, note that the strike train is locked and that a part of the rack hook lines up with the slot in the gathering pallet. This assures that the gathering pallet pin will not interfere with the rack teeth as the rack falls onto the snail before the strike occurs. Also note that the lock pin rests against the locking lever. As mentioned earlier, this means that the warning action takes care of itself and does not need to be set in the Schatz movement. Study Figure 91 for a "fine-tuning" adjustment that is needed in the gathering pallet position as the last note is struck. This adjustment enables the half hour mechanism to produce the required odd counts.

Figure 92 shows the positions of the two hammer tails when the strike train is at rest. They are resting between two of the ten arms of the hammer

FIG. 90. SCHATZ SHIP'S BELL STRIKE, WITH SNAIL REMOVED TO SHOW THE STAR CAM AND CANNON PINION.

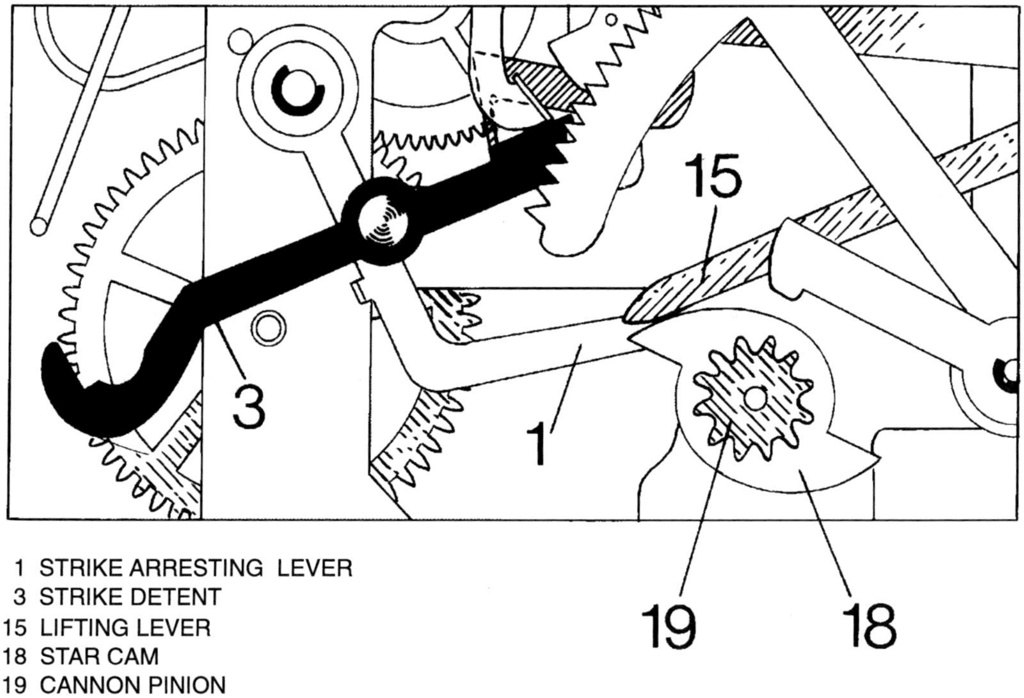

1 STRIKE ARRESTING LEVER
3 STRIKE DETENT
15 LIFTING LEVER
18 STAR CAM
19 CANNON PINION

FIG. 91. THE SCHATZ MECHANISM FOR COUNTING ODD AND EVEN BELLS

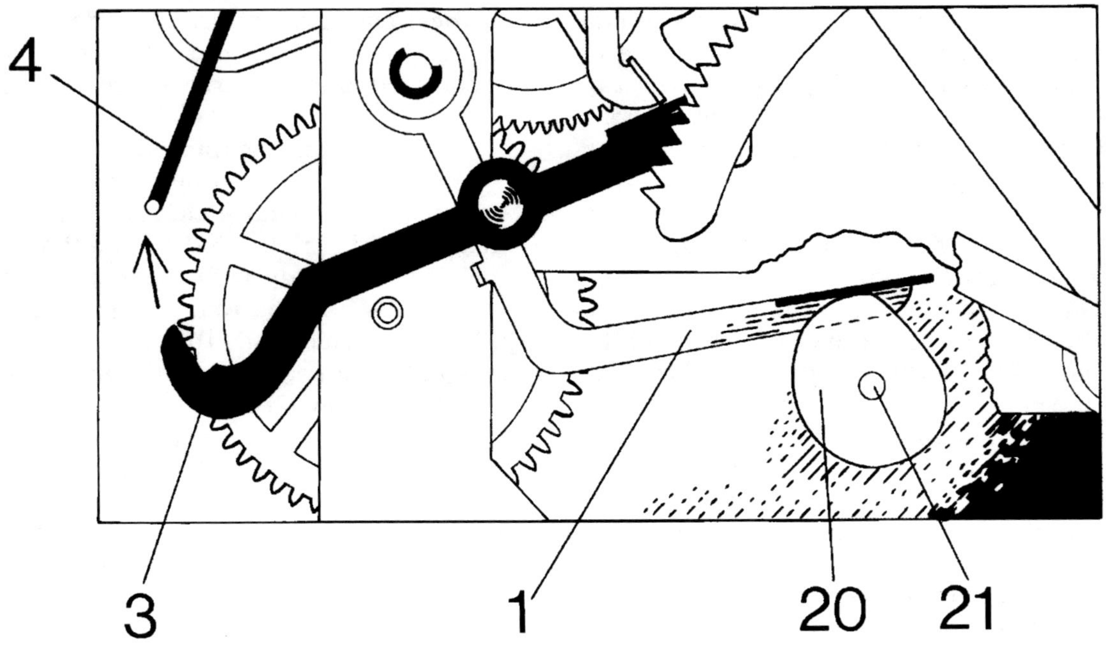

1 STRIKE ARRESTING LEVER
3 STRIKE DETENT
4 HAMMER LEVER EXTENSION
20 STRIKE ARRESTING CAM
21 CENTER ARBOR

TWO VIEWS OF THE GATHERING PALLET AND RACK HOOK AT THE INSTANT THE LAST BELL IS STRUCK. FOR THE MECHANISM TO WORK ON THE HALF HOUR (ODD) COUNTS, SET UP THE PARTS AS SHOWN.

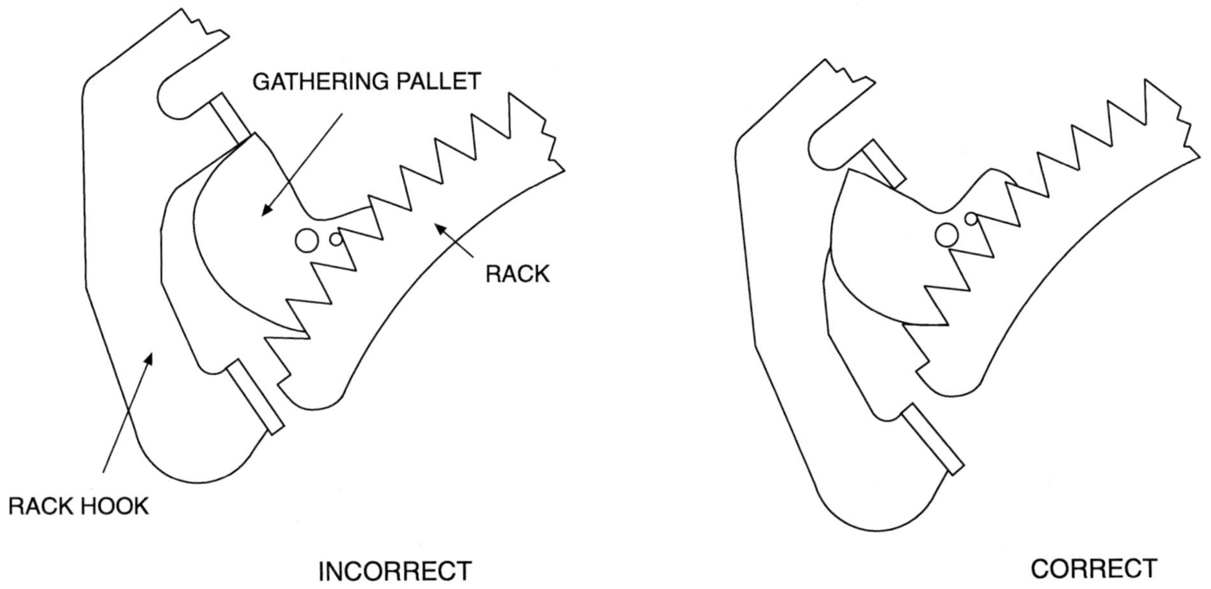

FIG. 92. REAR MOVEMENT STRIKE PARTS, SCHATZ SHIP'S BELL

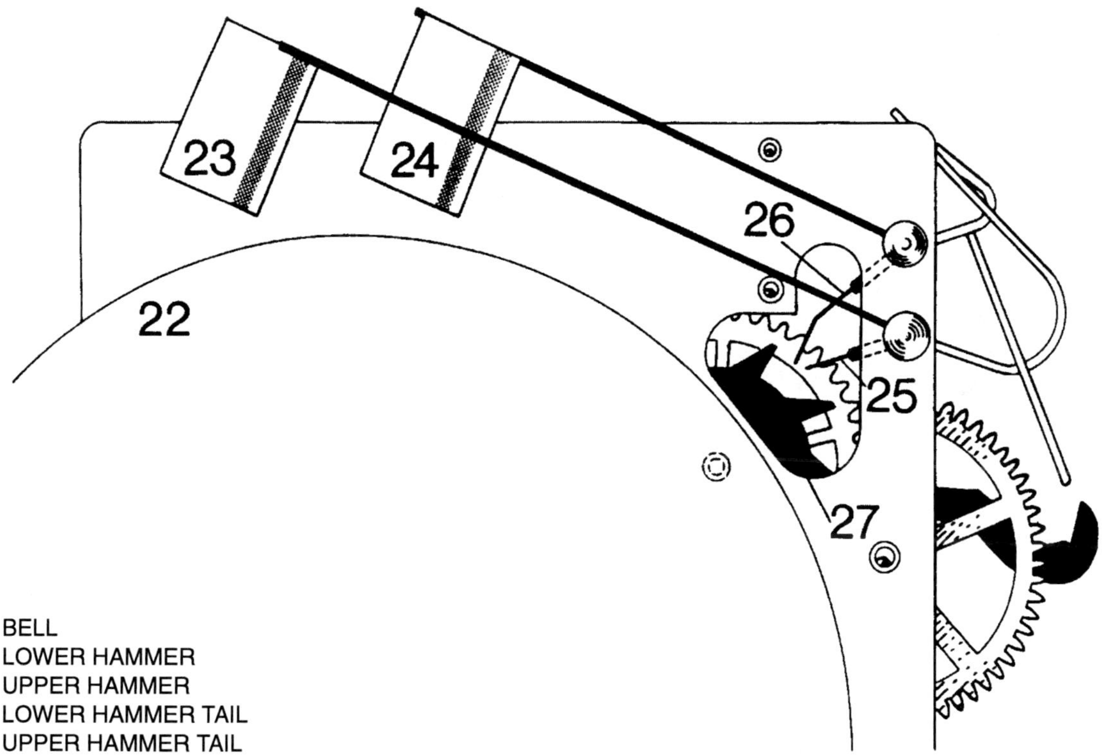

22 BELL
23 LOWER HAMMER
24 UPPER HAMMER
25 LOWER HAMMER TAIL
26 UPPER HAMMER TAIL
27 HAMMER LIFTING STAR

lifting star. This setup guarantees that the notes will come in pairs as they should. An *incorrect* setup would produce a single note first, followed by a pair of notes.

The snail, shown in Figure 89, must be installed the proper way. To check it, turn the minute hand as you watch the strike arresting lever. When the strike arresting cam raises the lever, you know you are approaching an odd (half) hour. Adjust the snail so the rack tail will fall upon the right side portion of the step. Check to be sure that the rack tail will fall upon the left part of the same step at the next hour.

Another important thing to check is the strike detent. It must securely catch and hold the hammer lever extension to suppress the last strike at each half hour. If it does not, look for a bent hammer lever extension or a strike arresting lever which does not ride up on its cam before the hour. I

Listen carefully to the sounds of the two hammers. They should be adjusted so they have the same clearance to the bell and strike it with equal force. The two hammers should sound the same; they should not sound like two different hammers hitting a bell. Instead, a listener should think there is just one hammer striking pairs of notes.

Finish the repair job with a test run. Each repairer will have his or her own ideas about the length of a test. In the case of the Schatz eight-day ship's bell clock, it seems reasonable to install the movement in the case and run it on the wall for two weeks (two winding intervals) without problems such as stopping, running fast, or running slow. In addition to keeping accurate time throughout a full week, the clock must strike the ship's bells properly throughout the full winding period. Try to catch any irregularity in striking, especially an even strike at the half hour. This indicates a problem, possibly intermittent, with the clock's odd (half hour) counting mechanism.

20

SALEM (MAUTHE) SHIP'S BELL

Ship's bell clocks bearing the name "Salem" contain a #W400 movement by the firm of F. Mauthe of Germany, which is no longer in business. The 3 inch diameter, seven jewel movement is also stamped "Salem Clock Co., Germany".

Figures 93 and 94 are front and rear photos of the movement. The distinctive ship's bell snail, with its three sectors of four steps each, is visible in the center of the front plate. It soon becomes apparent that most of the other strike parts are located on the back of this movement instead of on the front. A crescent shaped lower back plate allows removal of the mainspring barrels.

Figures 95 and 96 are drawings of the front and rear of the movement. The time and strike arbors (4) and (5) are reversed in this movement from the positions they occupy in most other movements; the time winding arbor is on the left, and the strike wind-

Fig. 93. Mauthe W400 movement, front view

Fig. 94. Rear view of the movement

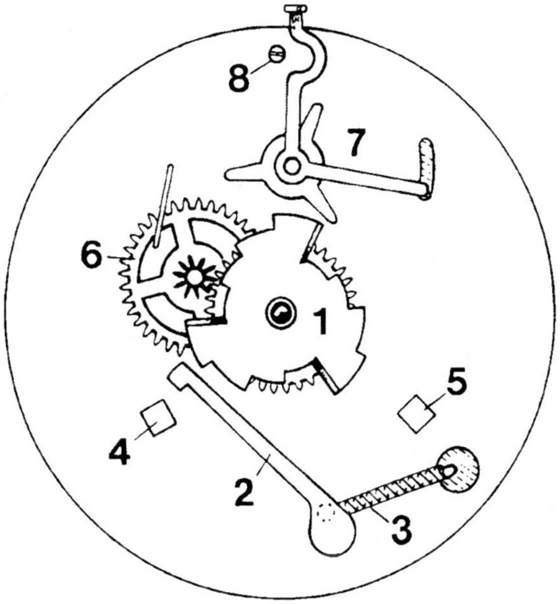

FIGURE 95. FRONT MOVEMENT PARTS, MAUTHE W400.

1 HOUR WHEEL & SNAIL
2 RACK TAIL
3 RACK SPRING
4 TIME WINDING ARBOR
5 STRIKE WINDING ARBOR
6 MINUTE WHEEL
7 STRIKE SILENCING MECHANISM
8 REGULATOR ARBOR

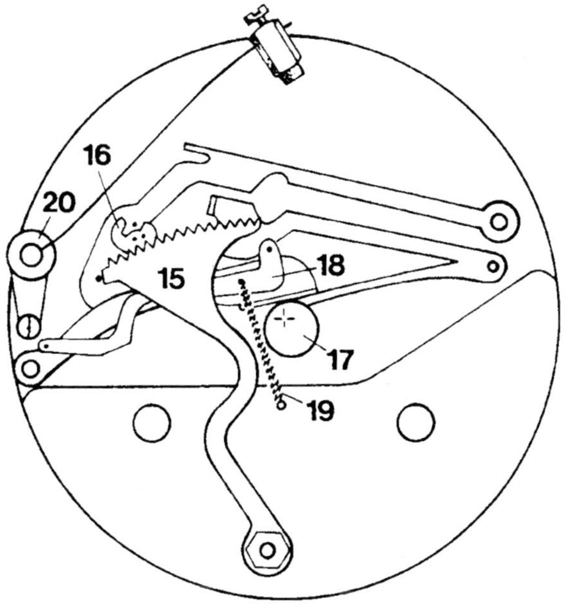

FIG. 96. THE REAR MOVEMENT PARTS, MAUTHE W400

15 RACK
16 GATHERING PALLET
17 HALF HOUR CAM
18 HALF HOUR ARRESTOR LEVER
19 ARRESTOR SPRING
20 HAMMER ASSEMBLY

ing arbor is on the right. The rack tail (2) and rack spring (3) are located on the front of the movement, but the rack is on the back plate. It is unusual for the rack and rack tail to be separated in this manner. The only other parts on the front plate are the motion work gears, including the minute wheel (6) and the hour wheel, underneath the snail; the regulator arbor (8); and the silencing mechanism (7). The silencing device is stiffened by a three-armed spring. Whenever the lever is moved to the "silent" position, the silencer lever hits an elongated tip on the fly blade to stop the strike train.

A mass of interlocking parts spans the back of the movement. These parts are all shown in Figure 96, and in Figures 97 and 98 they are separated into two groups to make them easier to see. The action starts with the star cam raising the lifting lever (13). This part also serves as the unlocking lever (14) and the warning lever (9). It pushes up the rack hook (11) to release the rack (15). The rack and gathering pallet (16) count off the strike notes with the total count determined by the snail (1) on the front plate. A pin wheel, not shown, lifts the hammer assembly (20). The six hammer lifting arms on this wheel are arranged as three pairs, giving the distinctive paired notes of the ship's bell strike. The strike train is stopped by the locking lever (10) at the end of the cycle.

Mechanism for Odd Counts (Half Hours)

The mechanism as explained so far would be able to strike only even counts. To make odd counts at the half hours, it is necessary to prevent the hammer from reaching the bell on the last stroke, although it does move. This is a good concept, since it means the last (even) note is still counted on the gear train, whether the audible count is odd or even. At the beginning of the next strike sequence, the pin wheel is ready to begin with a pair of notes. The important point is that the ship's bell strike should always begin with a pair of notes, never a single stroke (except at "one bell").

Figures 97, 98, and 99 show the Salem device for subtracting the last note on an odd count. The

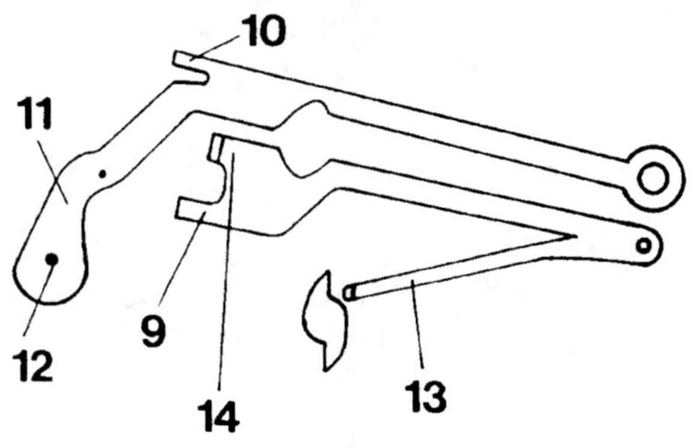

FIG. 97. BACK PLATE DETAIL.

9 WARNING LEVER
10 LOCKING LEVER
11 RACK HOOK
12 RACK PIN
13 LIFTING LEVER
14 UNLOCKING LEVER

hammer assembly (20) has a hammer stop pin (22) which pops up just in time to keep the hammer from hitting the bell. The way in which this is accomplished is curious. We start with the half hour cam (17) which raises the assembly shown in Figure 98 for a period of time spanning the half hour. The pin on the half hour arrestor (18) moves up toward the end of the rack. As the gathering pallet picks up the last tooth for the final pair of notes, the pin pushes against the end of the rack, using it as a fulcrum. Figure 98 shows that, as this happens, the lever pivots upward on the pivot point (21). The last hammer blow is caught on the downward stroke. Figure 99 illustrates the hammer assembly after it has been caught in this manner.

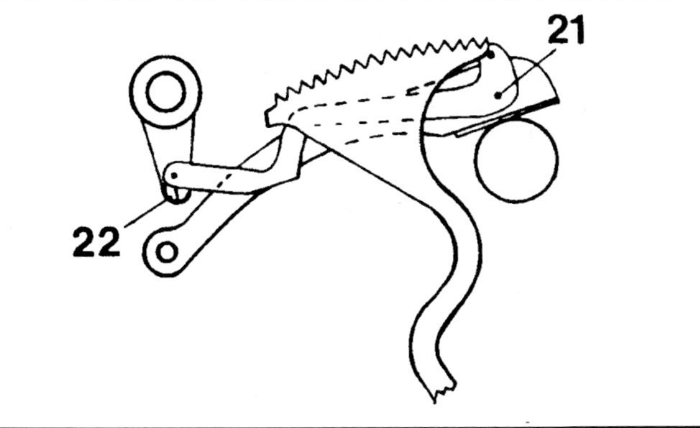

Fig. 98. Detail of levers following half hour strike: 21) pivot point for arrestor lever; 22) hammer stop pin.

During the third quarter, the half hour cam slowly allows the arrestor lever to return to the lower setting. The pin is no longer in a position to touch the rack at the hour. The hammer is held until the hour strike begins. As the pin wheel lifts the hammer, the arrestor lever falls away. The hammer strikes the bell. The last note is not subtracted, because the arrestor mechanism does not work at the hour.

This arrestor mechanism in the Salem ship's bell movement is an unusual device. Like most ship's bell clocks, the movement has a rack and snail strike and requires a method for preventing the hammer from hitting the bell the last time at the half hour. In other words, the second note of a pair is suppressed. This requires a device which operates quickly enough to catch the hammer. The motion is so swift that repairers find it difficult to understand what is happening. In fact, some may encounter a movement in which the device is not working, and

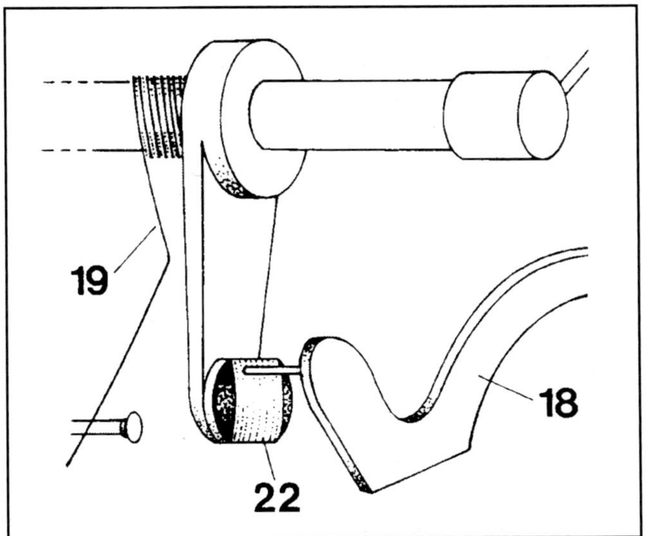

Fig. 99. Detail of hammer assembly: 18) half hour arrestor lever; 19) hammer spring; and 22) hammer stop pin.

as a result they do not know how the odd counts are to be produced. Comparing "odd count" mechanisms to other clockwork devices, they seem almost haphazard in design. The slightest bend in a lever makes the device inoperative. At least one repairer has reported a tendency for the Salem arrestor mechanism to fail when the strike mainspring is fully wound.

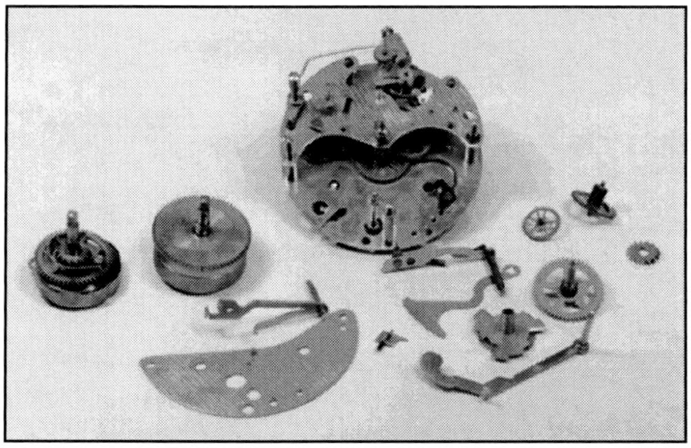

Fig. 100. Rear view of a partially disassembled Salem (Mauthe) movement.

Fig. 101. The strike mainspring is contained in a mainspring "box" (left) and the time mainspring is housed in a "going" barrel (right).

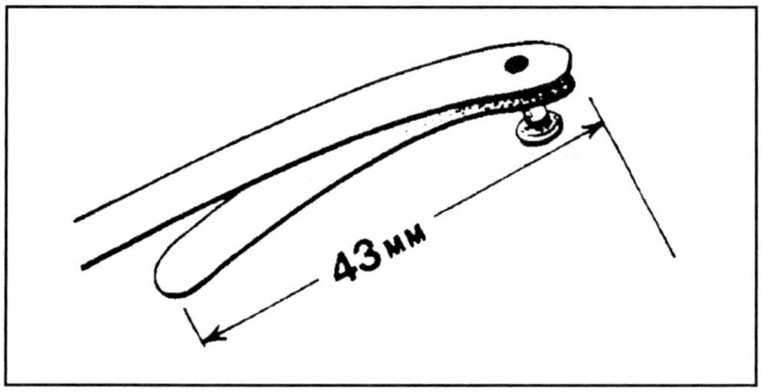

Fig. 102. The strike mainspring has a special end which hooks onto the wall of the mainspring "box".

Mainspring Data—Salem (Mauthe) Ship's Bell

Figure 100 shows a partially disassembled Salem ship's bell movement. The movement itself is seen in a rear view. Note the crescent shaped lower plate below and to the left of the movement. After the mainsprings were let down, this plate was removed, followed by the main wheels and mainsprings.

These two power units are very different from each other. The time train is powered by a "going barrel", which is shown on the right in Figure 101. This mainspring is 10mm wide by .010" thick by about 52" long. The spring ends are the "hole end" type. The strike mainspring is not housed in a going barrel. Instead there is a mainspring box, which is simply a container to surround the spring and provide a hooking point for the outer end. As shown in Figure 101 on the left, the strike main wheel is a miniature version of an American main wheel, with the click wheel staked onto the arbor and the click riveted to the wheel. The strike mainspring is .238" wide (approximately 1/4" or 6mm) by .009" thick by 48" long.

Figure 102 is a drawing of the outer end of the strike mainspring. It is a two-piece end which is stronger than a single section of mainspring. The rivet at the end of the spring hooks onto a slot in the wall of the mainspring box.

APPENDIX

TWO TOOLS FOR STRIKING CLOCKS

WIRE SPRING WINDER FOR STRIKE ARBORS
BENDING TOOL FOR STRIKE LEVERS

Considering all the tools and "repair tips" that have come to my attention over the years, two ideas stand out as the most useful for the repair and adjustment of striking mechanisms. Both are easy-to-make tools which will greatly improve and speed your work.

Wire Spring Winder for Strike Arbors

Striking clocks usually have wire springs to maintain tension on lifting levers, locking levers, and hammers. The repairer often finds these springs kinked or missing. If he or she winds a replacement directly onto the arbor by hand, the result can be unsightly. The wire is unruly and hard to handle, and it is difficult to control the tension when several wraps of wire intertwine and become tangled.

Fig. 103. Selection of brass spring wire: 30-gage (.010"); 26-gage (.016"); and 22-gage (.025")

There are several ways to wind neat, controlled wire springs. Machinists, for example, wind them in the lathe. There is also a method, described in W.R. Smith's book *Clockmaking and Modelmaking Tools & Techniques*, for winding springs with a steel crank held in a bench vise. The crank is clamped between two pieces of wood. The wire is wound around the crank, squeezed tightly between the wood surfaces. This produces closely wound coils whose diameter depends on the diameter of the steel rod used for the crank. This appendix describes an adaptation of that method applied specifically to strike arbors.

Think for a moment about the requirements for wire springs used in clock repairs. A spring should be wound with the coils relatively close together. This produces a neat appearance and helps to prevent the last coil from becoming wedged between the pivot shoulder and the movement plate. The spring fits freely over the arbor to allow for expansion and contraction of the coils as they flex. The ends of the spring should be hooked securely to the arbor and the plate or a movement pillar. Finally, the wire should have the amount of strength required for the clock.

Fig. 104. Winding a wire spring with the crank.

Figure 103 shows three spools of brass spring wire which cover the range of strengths needed for most clock work. The 30-gage wire is best for most strike arbors in American clocks, the 26-gage is stronger for hammer arbors in particular, and the 22-gage is needed only occasionally when an especially strong spring must be made. Be careful, especially when winding the stronger springs, since the spring can whip about as it is being wound up.

The winding setup is shown in Figure 104. A bench vise is used as a winding frame. Although a portable vise is shown for convenience in the photo, a heavier vise is better. A crank is held lightly in the vise jaws. As it is turned, the wire is wound up into a neat spring.

The crank is made from small diameter steel rod. A rod of a given diameter produces a spring of a somewhat larger diameter, since the wire is wound tightly around the crank and then springs back. As an example, a 5/32" diameter rod (about .157") produces a .200" I.D. spring coil with 26-gage wire. This is just right for American striking clock arbors, so the 5/32" diameter rod is suggested for the

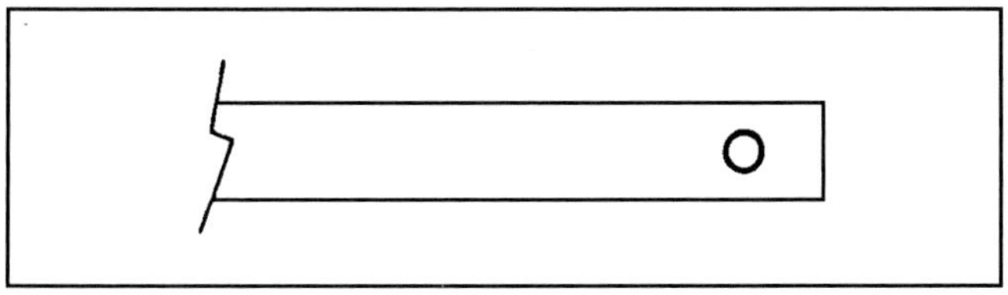

Fig. 105. Drill a 3/64" diameter hole 1/8" from the end.

crank. The other dimensions of the crank are not critical. Just make sure the straight spindle portion fits the vise you plan to use, as shown in the illustration. After forming the crank arm, drill a hole in the spindle portion to hold the end of the spring wire. Figure 105 shows that this is a radial hole. It is drilled near the end of the rod with a small drill such as 3/64". Remove any burrs from around the hole.

To get started winding a wire spring, cut a piece of brass spring wire six to eight inches long. Insert one end of the wire into the hole in the end of the crank spindle. Allow at least 1/4" of the wire to push through the other side of the hole. Bend the wire over the spindle, but do not kink it sharply. The purpose of the bend is to anchor the wire so that it stays hooked during winding. A sharp bend weakens the spring and is to be avoided.

Now look at the clock arbor and decide where the end of the new spring will be hooked. Typically, it is wrapped around a lever as shown in Figure 106. This end of the wire will be the end you just hooked through the spindle of the winding crank. Determine which way the wire must be coiled to make a spring that will move the clock arbor in the desired direction. This tells you which way to turn the crank. The penalty for turning the wrong way is small; just make a new spring wound the other way.

To begin winding, proceed as shown in Figure 104. Grasp the free end of the spring wire with pliers and pull. Working against this pulling action,

Fig. 106. New spring wound onto a Gilbert arbor

turn the crank to tightly wrap the wire around the spindle. Keep the pliers angled slightly toward the spindle end. This pressure will force the coil being formed to stay close to the previous one.

Make six or more wraps around the spindle, with the exact number determined by the length of spring you need. Leave at least an inch or two of unwound wire left. Carefully unhook the end of the spring wire from the hole in the spindle and remove it from the spindle. Slide the spring onto the clock arbor. Reform the bent end of the spring to grasp the hooking point on the arbor, which is probably a lever. See Figure 106. The other end of the spring should be hooked around a clock pillar or pushed through a hole in the plate, depending on the particular clock. Many old Seth Thomas movements have small brass posts for hooking the ends of the spring. Most American clocks do not have posts, and it takes a moment to decide exactly how and where it is best to hook the spring. Test the tension of the spring before forming the end which will hook and fasten it in place. If more tension is required, pull the spring tighter before hooking it.

In some cases, a strike arbor has a locking lever or a hammer lever very close to the end. This makes it impossible to simply slide the new spring past this obstacle and into position on the arbor. To overcome the problem, slide the spring onto the arbor and up against the lever. Thread the end of the spring around to the other side of the lever, then rotate the entire spring with a screwing motion. Continue until each one of the coils has gone to the desired side of the lever. This process might spread the coils slightly, but it still makes a neat job.

A Bending Tool for American Strike Levers

Pliers are not very precise tools for bending the soft steel wire strike levers found in old American clocks. These round wires are very easy to bend, but difficult to bend exactly as you want to bend them.

Figure 107 shows a typical American hammer arbor. The three pliable steel levers fastened to it

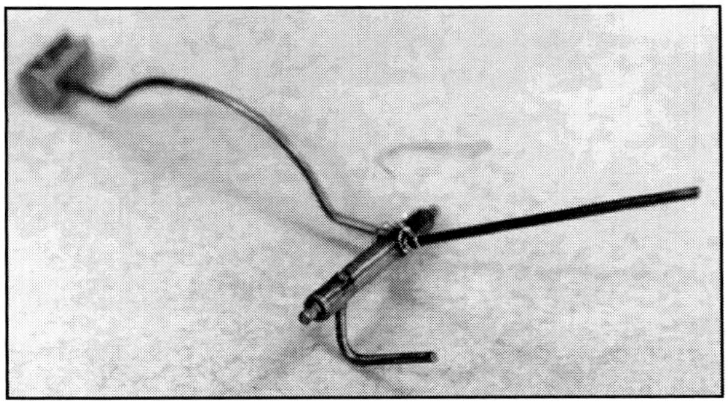

Fig. 107. A Gilbert hammer arbor with soft levers

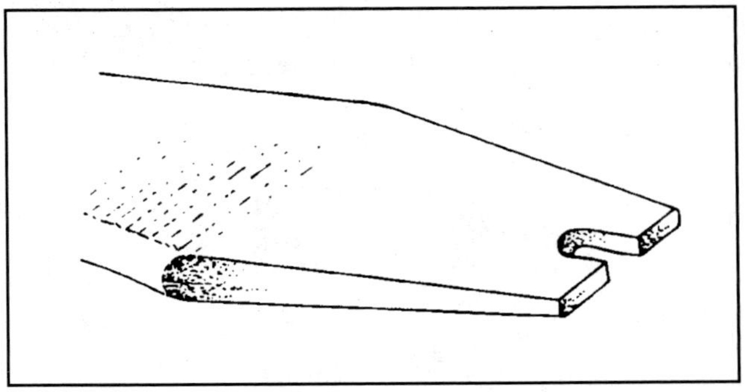

Fig. 108. A screwdriver blade is easily slotted to make a useful wire bending tool..

are similar to those in other American clocks. Count levers, drop levers, lifting levers, and other control wires of this type are often distorted by repairers. Even if unsightly bends and gouges from pliers are removed by a more careful repairer, there still remains the problem of shaping the wires to work properly. The simplest possible bending tool is shown in Figure 108. A slot is filed in the end of a screwdriver blade. The slot can be any convenient size, but a slot that is .070" wide and about the same depth is very useful on most clocks with soft steel levers. Inexpensive screwdrivers with soft blades can be filed. Those with hardened blades can be notched with a cutoff wheel held in a Dremel tool. If the edges of the slot are rounded to break the sharp corners, the bending tool is not as likely to mark the soft steel levers in the clock.

A single bending tool can be employed, or a pair can be used—one to hold the lever and the other to bend it in a precise manner. This tool makes it possible to make smaller radius bends neatly and accurately.